Brian Fleming Research & Learning Library
Ministry of Education
Ministry of Training, Colleges & Universities
900 Bay St. 13th Floor, Mowat Block
Toronto, ON M7A 1L2

COMMUNITY COLLEGES AND STEM

As United States policymakers and national leaders are increasing their attention to producing workers skilled in science, technology, engineering, and mathematics (STEM), community colleges are being called on to address persistence of minorities in these disciplines. In this important volume, contributors discuss the role of community colleges in facilitating access and success to racial and ethnic minority students in STEM. Chapters explore how community colleges can and do facilitate the STEM pipeline, as well as the experiences of these students in community college, including how psychological factors, developmental coursework, expertiential learning, and motivation affect student success. *Community Colleges and STEM* ultimately provides recommendations to help increase retention and persistence. This important book is a crucial resource for higher education institutions and community colleges as they work to advance success among racial and ethnic minorities in STEM education.

Robert T. Palmer is Assistant Professor of Student Affairs at the State University of New York—Binghamton, USA.

J. Luke Wood is Assistant Professor of Administration, Rehabilitation, and Postsecondary Education at San Diego State University, USA.

COMMUNITY COLLEGES AND STEM

Examining Underrepresented Racial and Ethnic Minorities

Edited by Robert T. Palmer and J. Luke Wood

Routledge
Taylor & Francis Group
NEW YORK AND LONDON

First published 2013
by Routledge
711 Third Avenue, New York, NY 10017

Simultaneously published in the UK
by Routledge
2 Park Square, Milton Park, Abingdon, Oxon OX14 4RN

Routledge is an imprint of the Taylor & Francis Group, an informa business

© 2013 Taylor & Francis

The right of the editors to be identified as the authors of the editorial material, and of the authors for their individual chapters, has been asserted in accordance with sections 77 and 78 of the Copyright, Designs and Patents Act 1988.

All rights reserved. No part of this book may be reprinted or reproduced or utilised in any form or by any electronic, mechanical, or other means, now known or hereafter invented, including photocopying and recording, or in any information storage or retrieval system, without permission in writing from the publishers.

Trademark notice: Product or corporate names may be trademarks or registered trademarks, and are used only for identification and explanation without intent to infringe.

Library of Congress Cataloging in Publication Data
Community colleges and STEM : examining underrepresented racial and ethnic minorities / edited by Robert T. Palmer and J. Luke Wood.
　pages cm
　Includes bibliographical references and index.
　1. Science—Study and teaching (Higher)—United States. 2. Technology—Study and teaching (Higher)—United States. 3. EngineeringStudy and teaching (Higher)—United States. 4. Mathematics—Study and teaching (Higher)—United States. 5. Science students—United States. 6. Community college students—United States. 7. Minority college students—United States. I. Palmer, Robert T., editor of compilation. II. Wood, J. Luke, 1982– editor of compilation.

　Q183.3.A1C653 2013
　507.1'073—dc23

ISBN: 978-0-415-82110-0 (hbk)
ISBN: 978-0-203-56844-6 (ebk)

Typeset in Bembo and Stone Sans
by EvS Communication Networx, Inc.

Printed and bound in the United States of America by Publishers Graphics, LLC on sustainably sourced paper.

CONTENTS

Foreword ix
Estela Mara Bensimon and Cecilia Santiago

Preface xiii

Acknowledgments xix

PART I
Pathways to Success: The Role of Community Colleges in Promoting Access to Minorities in STEM   **1**

  1 Community Colleges and Underrepresented Racial and Ethnic Minorities in STEM Education: A National Picture   3
      Xueli Wang

  2 The Impact of State Policy on Community College STEM Programs   17
      Pamela Eddy

  3 The Need for Integrated Workforce Development Systems to Broaden the Participation of Underrepresented Students in STEM-Related Fields   37
      Victor Hernandez-Gantes and Edward C. Fletcher Jr.

PART II
A New Dimension to the Discourse on Minority Students, STEM, and Community Colleges   57

4 An Expectancy-Value Model for the STEM Persistence of Ninth-Grade, Underrepresented Minority Students   59
Lori Andersen and Thomas J. Ward

5 The Effect of Non-Cognitive Predictors on Academic Integration Measures: A Multinomial Analysis of STEM Students of Color in the Community College   75
Marissa Vasquez Urias, Royel M. Johnson, and J. Luke Wood

6 STEMming the Tide: Psychological Factors Influencing Racial and Ethnic Minority Students' Success in STEM at Community Colleges   91
Terrell L. Strayhorn, Michael Steven Williams, Derrick L. Tillman-Kelly, and Marjorie Dorimé-Williams

7 The Propensity to Avoid Developmental Math in Community College: A Focus on Minority Students   101
Bobbie Everett Frye. James E. Bartlett, II, and Kelly D. Smith

8 Moving Beyond the Barrier of Mathematics and Engaging Culturally Relevant Pedagogy in the Classroom for Racial and Ethnic Minority STEM Students in Community Colleges   123
Denise Yull

PART III
Examining the Experiences of Minority Students in Community Colleges: Diverse Contexts   139

9 Minority Serving Community Colleges and the Production of STEM Associate's Degrees   141
Frances King Stage, Ginelle John, Valerie C. Lundy-Wagner, and Katherine Mary Conway

10 Creating Successful Pathways for Asian Americans and Pacific Islander Community College Students (AAPIs) in STEM   156
Dina C. Maramba

11 Constraints and Opportunities for Practitioner Agency in STEM Programs in Hispanic Serving Community Colleges   172
Megan M. Chase, Estela Mara Bensimon, Linda Taing Shieh, Tiffany Jones, and Alicia C. Dowd

12 Achieving Success: A Model of Success for Black Males in STEM at Community Colleges 193
    Robert T. Palmer and Zachary M. DuBord

About the Editors   209
About the Contributors   211
Index   218

FOREWORD

Estela Mara Bensimon and Cecilia Santiago

Community colleges are the primary point of entry into higher education for the majority of underrepresented college students, many of whom hope to transfer to four-year colleges and earn a bachelor's degree, and, in some cases, go on to earn a graduate or professional degree. As an institution, the community college has been a place of hope and possibility for youth and adults who are less affluent or who have been deprived of the academic resources and experiences that ease the transition to higher education. From the standpoint of access, community colleges have been remarkably successful, so much so that today a greater proportion of first-time students start out at a community college than at a four-year college.

However, access alone is not an adequate measure of equity in educational opportunity. Community colleges are being challenged by philanthropic organizations, state governments, the president of the United States, and the public to produce more degrees, certificates, and students who are academically prepared to transfer to four-year colleges. But more is needed than a larger number of degrees bestowed. Several national reports and blue ribbon panels have placed a *call to action* for higher education institutions to broaden the participation of underrepresented students in STEM. Increasing the representation of students of color in STEM fields is vital to the continued scientific advancement of the United States. The sheer size of minority students in community colleges and the impending mass retirement of America's older, relatively well-educated generation of baby boomers (Kelly, 2008) means that the U.S. science and engineering workforce will become more reliant on the participation of racial and ethnic groups that STEM fields have not served equitably. Given the large concentration of students of color in community colleges, particularly in large urban areas such as Los Angeles, New York City, Boston, Chicago, Milwaukee,

San Francisco, Seattle, and hundreds of smaller predominantly minority segregated cities, community colleges are vital to fulfilling the national agenda for more equitable and successful minority access and graduation in STEM.

In a statement on broadening participation in STEM before the House Subcommittee on Research and Science Education, Alicia Dowd pointed out that "For too long, our approach to improving diversity in STEM has been overly focused on the 'demand' side of the problem, or 'fixing' presumed student deficits through attempts to improve their aspirations, motivation, or willingness to succeed. Indeed a great many of the efforts to increase the entry of underrepresented students in community college students into STEM fields represent what organizational learning theorists" (Argyris, 1993; Argyris & Schön, 1996) describe as a functional "fix" (e.g., articulation agreements, special programs) that does not question the underlying principles that created the problem in the first place. In other words, the focus is on problem-solving rather than problem-questioning of the value systems in which the process or problem is embedded (Witham & Bensimon, 2012). Without questioning the underlying conditions that contribute to the underrepresentation of racial minority groups in the first place, significant improvements in outcomes are unlikely.

This book provides practitioners, policymakers, and scholars with information and strategies that address how community colleges can more intentionally assume responsibility for analyzing the conditions that stand in the way of more successful participation and outcomes of minority students in STEM. This is an important contribution to the national agenda for greater equity and inclusiveness in STEM because it highlights the critical role community colleges can play. As many of the book's authors assert, community colleges have not always been considered or perceived as a viable solution to the leak in the STEM pipeline, despite the disproportionate enrollment of students of color at these institutions.

As you immerse yourself in the rich data and recommendations this book presents, we urge readers to consider a recommendation that is often left out of the equation—to make sustainable progress in closing the opportunity and achievement gap for underrepresented students in STEM, it is important to consider a theory of change that is founded on the principles of democracy and social justice, and not solely driven by an economic agenda.

References

Argyris, C. (1993). *Knowledge for action: A guide to overcoming barriers to organizational change.* San Francisco, CA: Jossey-Bass.

Argyris, C., & Schön, D. A. (1996). *Organizational learning II: Theory, method and practice.* Reading, MA: Addison-Wesley.

Dowd, A. C. (2010, March 16). Broadening Participation in STEM. Testimony before the House Subcommittee on Research and Science Education of the Science and Technology Committee. Washington, DC

Kelly, P. (2008). *Beyond social justice: The threat of inequality to workforce development in the western United States*. Boulder, CO: Western Interstate Commissions for Higher Education.

Witham, K. A., & Bensimon, E. M. (2012). Creating a culture of inquiry around equity and student success. In S. D. Museus & U. M. Jayakumar (Eds.), *Creating campus cultures: Fostering success among racially diverse student populations* (pp. 46–67). New York, NY: Routledge.

PREFACE

Interest-Convergence Striking While the Iron Is Hot

Scientific advancements have led to America's status as a world leader, accounting for nearly half of U.S. economic development in the past 50 years. However, the United States is at a critical juncture as trends in achievement, degree production, and global competitiveness signal decline in the nation's dominance. A 2007 report from the U.S. Department of Labor illustrates several interrelated trends that are cause for concern. First, inadequate academic preparation in preK-12 education (largely attributed to ineffective instruction) has led to low levels of STEM readiness. For example, the United States (once a global leader) now ranks 28th in math literacy and 24th in science competency (Kuenzi, 2008). Second, low proportions of U.S. college-going students are pursuing degrees in science and math in comparison to other nations. This issue is corroborated by data from the National Science Foundation which indicates that the United States is ranked 20th among other nations in the proportion in degree earners in natural sciences and engineering (Kuenzi, 2008). Partially, this is due to a lack of role models in STEM (particularly for students of color), skyrocketing tuition and fees, and unsatisfactory experiences in STEM courses (American Association of State Colleges and Universities, 2005). Third, there are evident disparities in the enrollment and success of historically underrepresented and underserved students, including women, students of color, and low-income students. These students are significantly less likely than their male, White, Asian, and affluent student peers to enter into and succeed in STEM fields (U.S. Department of Labor, 2007).

In order to maintain its prowess as a global leader in science and technology, the United States must simultaneously improve the achievement of students in STEM (preK-20) while increasing the proportion of collegians entering STEM majors and, as a result, STEM careers. In recent years, such efforts, particularly federal interventions and dollars, have focused on students of color. Recent attention on the community college's role in the STEM pipeline brings to bear the notion of interest convergence. As proffered by Derrick Bell (1980), "the principle of interest convergence provides: The interest of blacks in achieving racial equality will be accommodated only when it converges with the interests of whites" (p. 523). In essence, interest convergence suggests that advances for Black (and other minority) groups only occur when their interests align with those in power.

Students of color have been largely overlooked in STEM fields for decades, but now, have increasing importance in STEM conversations. Business leaders and policy makers have begun to realize that the United States cannot continue to compete in a world marked by scientific and technological advancement without the incorporation of students of color into the STEM pipeline. However, the editors of this volume assert, as do several of our colleagues (see Baber, 2012; Essien-Wood, 2010), that the impetus for the recent national focus on students of color in STEM is purely motivated by national interests. Evidence of this assertion is seen in a 2005 testimony (and subsequent report) given to the House Committee on Education and the Workforce by Cornelia M. Ashby, then Director of Education, Workforce, and Income Security Issues. In her testimony, she stated:

> The United States is a world leader in scientific and technological innovation. To help maintain this advantage, the federal government has spent billions of dollars on education programs in the science, technology, engineering, and mathematics (STEM) fields for many years. However, *concerns have been raised about the nation's ability to maintain its global technological competitive advantage in the future* ... the estimated number of women, African-Americans, and Hispanic-Americans employed in STEM fields increased, women and minorities remained underrepresented relative to their numbers in the civilian labor force.... To the extent that *these populations have been historically underrepresented in STEM fields, they provide a yet untapped source* of STEM participation in the future. (emphasis added)

The message communicated here is clear; the recent focus on students of color is one of national interest and necessity, not necessarily one of social justice or a perceived ethical imperative. Bearing this in mind, scholars and practitioners concerned about the nation's continued competitiveness in a global-market economy as well as the ethical and social implications for diversification must strike while the iron is hot. In essence, the current focus on students of color in STEM must be seen as an opportunity. It is an opportunity to change systems, structures, programs, policies, and practices that have prevented the incorporation of historically underrepresented and underserved communities from entering into and succeeding in college (and particularly in STEM).

This brings us to the focus of this book on students of color in STEM in community colleges. Community colleges are an increasingly important site for STEM interventions given that they enroll high proportions of students of color. In fact, the overwhelming majority of students of color begin their academic careers in community colleges. This book seeks to capitalize on current trends by providing a platform for cutting edge research that serves to demarcate the factors impacting the success of STEM students of color in the community college. It is our hope that the chapters presented serve to advance three primary foci: (a) to advance discourse among researchers about enrollment, persistence, and achievement disparities in STEM; (b) to better understand the role of academic, non-cognitive, environmental, social, and institutional factors on students' of color success in STEM; and (c) to further discussions on promising practices that ultimately bolster the community college's ability to graduate and transfer underrepresented minorities in STEM.

Overview of Chapters

This book is divided into three parts and consists of 12 chapters. The first part examines the role of community colleges in promoting access to minorities in science, technology, engineering, and mathematics (STEM). Chapter 1, "Community Colleges and Underrepresented Racial and Ethnic Minorities in STEM Education: A National Picture," by Xueli Wang presents the role of community colleges in broadening the educational pipeline in STEM, especially for underrepresented minorities. Wang also discusses research and policy implications surrounding the intersection between community colleges and STEM education, with a focus on underrepresented minorities.

Chapter 2, "The Impact of State Policy on Community College STEM Programs," by Pamela Eddy studies the influence of a policy in Virginia that is focused on increasing pursuit of STEM degrees in high-demand areas, with the objective of fueling economic development. This chapter provides a review of the strategic plan of the Virginia Community College System, discusses the beginning impact of this new policy, and provides a framework for replication in other states.

In Chapter 3, "The Need for Integrated Workforce Development Systems to Broaden the Participation of Underrepresented Students in STEM-related Fields," Victor Hernandez-Gantes and Edward C. Fletcher Jr. analyze literature supporting the call for the integration of workforce education systems as a means to broaden participation in STEM related fields. The authors provide an overview of participation and degree completion trends, with particular focus on underrepresented groups and specific STEM fields. Subsequently, the chapter examines current initiatives seeking to broaden the participation of underrepresented groups in STEM career paths. Finally, this chapter explores the implications for an integrated workforce education and development system to engage communities and education entities in coherent efforts to broaden the participation of underrepresented groups in STEM education and work.

The second part of the book focuses on providing new dimension to the discourse on racial and minority students, STEM, and community colleges. In Chapter 4, "An Expectancy-Value Model for the STEM Persistence of Ninth Grade Underrepresented Minority Students," Lori Andersen and Thomas J. Ward used data from the High School Longitudinal Study of 2009 to examine factors that predict plans to persist in STEM for a nationally representative sample of underrepresented minority students and found that females were more than twice as likely as males to plan to persist. Incongruences between minority students' identities and STEM identities are barriers to motivation and persistence. This chapter concludes by discussing methods that community colleges can employ to increase the participation of racial and ethnic minority students in STEM, such as the provision of role models, career planning assistance, and the use of culturally responsive instruction.

Marissa Vasquez Urias, Royel M. Johnson, and J. Luke Wood examined non-cognitive predictors of academic integration for students of color in STEM in Chapter 5, "The Effect of Non-Cognitive Predictors on Academic Integration Measures: A Multinomial Analysis of STEM Students of Color in the Community College." Their research highlights the important role that an internal locus of control can have on integration. They also articulated why college professionals may need to reinforce the importance of faculty-student interactions, meeting with academic advisors, and using the library, to greater levels of math self-efficacy. In all, findings from this study show that self-efficacy, degree utility, locus of control, and action control can have nuanced effects on integration for STEM students of color. In Chapter 6, "STEMming the Tide: Psychological Factors Influencing Racial and Ethnic Minority Students' Success in STEM at Community Colleges," Terrell L. Strayhorn, Michael Steven Williams, Derrick L. Tillman-Kelly, and Marjorie Dorimé-Williams analyzed data from a national sample of students using the Community College Student Experiences Questionnaire to understand the influence of background and social psychological factors on minority students' success in STEM at community colleges. This chapter concludes with implications for practice and research focused on advancing the success of minority students in STEM at the community college level.

Chapter 7, "The Propensity to Avoid Developmental Math in Community College: A Focus on Minority Students," by Bobbie Everett Frye, James E. Bartlett, II, and Kelly D. Smith, investigates whether there is a difference in outcomes (grade point average, credits earned, credentials earned, transfer, persistence) between two groups: (a) an avoider (control) group of minority students who avoid developmental math coursework in the first year of enrollment and (b) a non-avoider (treatment) group of minority students who attempt developmental math coursework in the first year of enrollment. The results suggested completion of developmental math credits led to successful completion of college level math. The chapter concludes with implications for practical suggestions and considerations.

Denise Yull discusses effective strategies for teaching racial and ethnic minority students in STEM in Chapter 8, "Moving Beyond the Barrier of Mathematics and Engaging Culturally Relevant Pedagogy in the Classroom for Racial and Ethnic Minority STEM Students in Community Colleges." Yull draws upon her experiences with pedagogical strategies that she found to be impactful in improving the educational experience and success of minority students pursuing STEM in community colleges. The chapter concludes with implications for practitioners and educators in community colleges.

In Chapter 9, "Minority Serving Community Colleges and the Production of STEM Associate Degrees," Frances King Stage, Ginelle John, Valerie C. Lundy-Wagner, and Katherine Mary Conway add to existing research on MSIs and the minority STEM pipeline by highlighting STEM and non-STEM completions at two-year colleges by racial/ethnic groups at MSIs and non-MSIs. The chapter concludes with implications regarding the role that MSI community colleges can play in increasing achievement among minority students in STEM.

In Chapter 10, "Creating Successful Pathways for Asian Americans and Pacifier Islander Community College Students (AAPIs) in STEM," Dina C. Maramba focuses on how AAPI community college students fit in the discourse of underrepresented minority students and STEM. This chapter is comprised of several sections, which provide context of the complexity of the AAPI category, contextualize AAPIs in community colleges, provide insight into AAPIs' participation in STEM, and describe current initiatives to help increase access and success among AAPI STEM students in community colleges. The chapter concludes with implications for researchers, practitioners, and policy makers.

Megan M. Chase, Estela Mara Bensimon, Linda Taing Shieh, Tiffany Jones, and Alicia C. Dowd discuss how institutional agents can be used to facilitate the success of Latino/as in STEM in Chapter 11, "Constraints and Opportunities for Practitioner Agency in STEM Programs in Hispanic Serving Community Colleges." Data in this chapter emerged from a larger study which focused on what motivated institutional agents to use their resources to facilitate success in STEM among Latinos/as. The chapter looks at how institutional agents in community colleges can work collectively or separately to be intentional about advancing the participation and success of Latino/a students in STEM.

In the final chapter, "Achieving Success: A Model of Success for Black Males in STEM at Community Colleges," Robert T. Palmer and Zachary M. DuBord draw from extant literature on Black males, STEM, and community colleges to construct a model that institutional agents can use to increase the participation and success of Black males in community colleges pursing opportunities in STEM.

References

American Association of State Colleges and Universities. (2005). Strengthening the science and mathematics pipeline for a better America. *Policy Matters, 2*(11), 1–4.

Ashby, C. M. (2005). *Higher education: Science, technology, engineering, and mathematics trends and the role of federal programs. Testimony before the Committee on Education and the Workforce, House of Representatives.* Washington, DC: Government Accountability Office. Retrieved from http://www.gao.gov/new.items/d06702t.pdf

Baber, L. D. (2012). *The interest convergence dilemma in STEM education: Project STEP-UP.* Champaign: University of Illinois at Urbana-Champaign.

Bell, D. (1980). Brown and the interest-convergence dilemma. In D. Bell (Ed.), *Shades of Brown: New perspectives on school desegregation* (pp. 90–106). New York, NY: Teachers College.

Essien-Wood, I. (2010). *Undergraduate African American females in the sciences: A qualitative study of student experiences affecting academic success and persistence* (Unpublished doctoral dissertation). Arizona State University Tempe.

Kuenzi, J. J. (2008). *Science, technology, engineering, and mathematics (STEM) education: Background, federal policy, and legislative action. Report for Members and Committees of Congress.* Washington, DC: Congressional Research Service.

U.S. Department of Labor. (2007). *The STEM workforce challenge: The role of the public workforce system in a national solution for a competitive science, technology, engineering, and mathematics (STEM) workforce.* Washington, DC: Author.

ACKNOWLEDGMENTS

The editors would like to acknowledge the numerous contributors to this volume, including (in order of appearance): Xueli Wang, University of Wisconsin-Madison; Pamela Eddy, The College of William and Mary; Victor Hernandez-Gantes and Edward C. Fletcher Jr., University of South Florida; Lori Andersen and Thomas J. Ward, College of William and Mary; Marissa Vasquez Urias, San Diego State University; Royel M. Johnson, University of Illinois at Urbana-Champaign; Terrell L. Strayhorn, Michael Steven Williams, Derrick L. Tillman-Kelly, and Marjorie Dorimé-Williams, The Ohio State University; Bobbie Everett Frye, Central Piedmont Community College, James E. Bartlett, North Carolina State University, and Kelly D. Smith, Central Piedmont Community College; Denise Yull, SUNY Binghamton; Frances King Stage, Ginelle John, and Valerie C. Lundy-Wagner, New York University; Katherine Mary Conway, CUNY–Borough of Manhattan Community College; Dina C. Maramba, SUNY Binghamton; Megan M. Chase, Estela Mara Bensimon, Linda Shieh, Tiffany Jones, and Alicia C. Dowd, University of Southern California; and Zachary M. DuBord, SUNY Binghamton. In addition to the chapter contributors, we would also like to thank Estela Bensimon and Cecilia Santiago at the University of Southern California for writing the foreword for this volume.

Robert T. Palmer would like to acknowledge J. Luke Wood: This has been the second book that we have completed together. I love working with someone as talented and motivated as you!

J. Luke Wood would like to thank Robert T. Palmer for his leadership in completing this important volume. I truly appreciate your work ethic and passion for issues of equity and social justice in education.

This book is dedicated to all the underrepresented racial and ethnic minority students who are striving for success in STEM and beyond. If you can conceive it, you can achieve it!

PART I

Pathways to Success

The Role of Community Colleges
in Promoting Access to Minorities in STEM

1

COMMUNITY COLLEGES AND UNDERREPRESENTED RACIAL AND ETHNIC MINORITIES IN STEM EDUCATION

A National Picture

Xueli Wang

As the demand for graduates in science, technology, engineering, and mathematics (STEM) disciplines continues to grow at a rapid rate, participation of traditionally underrepresented racial and ethnic minorities in these critical fields of postsecondary study still presents cause for concern (e.g., Anderson & Kim, 2006; National Academies 2005 "Rising Above the Gathering Storm" Committee, 2010; Smyth & McArdle, 2004). The discussion on how to promote the representation and success of these students has certainly led to numerous empirical studies and policy reports over the past few decades.

Educating nearly one-half of beginning college freshmen, community colleges are integral to achieving the Obama administration's goal of increasing the proportion of adults ages 25–35 with postsecondary degrees and workplace credentials from 40% to 60% by 2020 (Obama, 2009). For many members of underrepresented racial and ethnic minority (URM) populations, community colleges serve as an entry point to postsecondary education (e.g., Bailey et al., 2004; Dowd, 2007; Wang, 2009) and offer a unique opportunity in the preparation of a future STEM workforce that reflects the diversity of the U.S. population (Dowd, 2011). Therefore, in addressing national concerns about the STEM pipeline, much of the effort will rely on the nation's over 1,000 community colleges, which, due to their large and diverse enrollments, play a chief role in building the STEM workforce and widening access to and success in college settings for youth and returning adults.

However, it was not until recent years that the community college started to garner well-deserved attention in helping resolve the issue of underrepresentation of racial and ethnic minority students in the STEM education pipeline, a critical issue of national importance. As a result, although there has been a heightened recognition of the pivotal role of community colleges in facilitating these students' STEM educational participation and success, much of the discussion is based on policy relevance. Very limited descriptive and empirical data exist to closely capture the role of community colleges in serving underrepresented racial and ethnic minority students in STEM.

Therefore, this chapter presents nationally representative statistics that describe the extent to which community colleges are serving underrepresented racial and ethnic minorities in STEM. Two national longitudinal studies sponsored by the Institute of Education Sciences (IES) provide data for this analysis: the Education Longitudinal Study of 2002 (ELS:2002) and the 2004/09 Beginning Postsecondary Students Longitudinal Study (BPS:04/09). With ELS:2002 following a particular high school cohort and BPS:04/09 focusing on all beginning postsecondary college students, analyses presented in this chapter examine how community colleges educate both traditional and non-traditional age racial and ethnic minority students in STEM fields in relatively recent years.

The analyses begin with an ELS:2002-based descriptive profile of recent high school students entering community colleges who aspire to and enroll in STEM fields of study, followed by a BPS:04/09-based examination of all beginning postsecondary students enrolled in STEM at community colleges, those who transferred into STEM fields at four-year institutions, and beginning four-year students in STEM fields who enrolled at community colleges. To provide the general context, STEM enrollment patterns of underrepresented racial and ethnic minorities will be contrasted against White students and Asian American students[1] who traditionally are well represented in postsecondary education and particularly in STEM disciplines (Anderson & Kim, 2006; Chubin, May, & Babco, 2005; Hurtado et al., 2007; May & Chubin, 2003). These analyses are accompanied by discussions situated within the existing research and policy literature. The chapter concludes with implications for future research and policy aimed at promoting the role of community colleges in moving underrepresented racial and ethnic minorities forward along the STEM educational pipeline.

Recent High School Graduates Starting at Community Colleges: STEM Aspirations and Entrance

To better understand the role of community colleges in expanding the STEM education pipeline for underrepresented racial and ethnic minorities, it is important to begin with an understanding of STEM-related aspirations and academic choices among high school graduates seeking postsecondary entry through community colleges. Despite this need, relevant analyses remain surprisingly scarce. For community colleges to better serve underrepresented racial and ethnic minority students pursuing STEM fields, a basic but often neglected question is how many of these students community colleges are recruiting and educating? Among students who are interested in and who actually enroll in

STEM fields at two-year colleges, how many are underrepresented racial/ethnic minorities?

To answer these questions, the Education Longitudinal Study of 2002 (ELS:2002) represents a good source of data. ELS:2002 is a national, longitudinal survey conducted by the National Center for Education Statistics (NCES) of the Institute of Education Sciences (IES), U.S. Department of Education, and in collaboration with several research institutes. ELS:2002 collected information on survey participants' educational experience in high school and postsecondary education, as well as their transition to and success in postsecondary education and the workforce. Drawing upon a longitudinal design, ELS:2002's baseline survey was completed in 2002, when participants were high school sophomores. Then, the first follow-up survey was conducted in 2004, and participants' high school transcript data were added to the database. The first follow-up survey also includes participants' demographic data. The second follow-up survey was completed in 2006, effectively two years after high school graduation for most survey participants. New variables were added to the database, such as individuals' postsecondary enrollment and experiences, social and economic returns of education, and newly acquired adult roles. For more information on ELS:2002, see http://nces.ed.gov/surveys/els2002.

The sample used in this analysis included students who participated in both the first and second follow-up interviews of ELS:2002 and who had enrolled in a community college by 2006 (approximately 3,470 students, accounting for roughly 35% of all postsecondary participants by 2006). All analyses were weighted using the appropriate panel weight (F2F1WT) and therefore the results generalize to the population of the spring 2004 high school graduates who attended postsecondary education through a community college within two years of high school graduation.

Based on the ELS:2002 data, roughly 16% of the 2004 high school graduates enrolled in postsecondary education by 2006 aspired to enter a STEM field of study and about 12% of the same high school cohort declared a STEM major by 2006. Included in the STEM-aspiring students are about 21% of the 2004 ELS high school cohort enrolled at public four-year institutions, 17% of the recent high school graduates at private four-year institutions, and 10% out of all 2004 high school graduates at community colleges. In terms of this 2004 cohort who actually declared STEM majors by 2006, 17% of the cohort members enrolled at public four-year institutions did so, 15% at private four-year institutions, and a little over 7% at community colleges. Taking this general profile as a point of departure, Table 1.1 presents racial/ethnic percentage summaries of STEM aspirations and entrance of high school graduates of 2004 who enrolled in a community college by 2006, in contrast to their counterparts at public and private four-year institutions. Note that given the use of a national sample with clustering and sampling weight features, the actual sample size of each category is less informative than the percentage. Therefore, only weighted percentages are reported.

TABLE 1.1 Percentage of High School Graduates of 2004 Who Aspired[1] to Enter STEM Fields and Who Enrolled[2] in STEM Fields by 2006

Race/Ethnicity	Community College		Four-Year Public		Four-Year Private	
	STEM aspirants	STEM entrants	STEM aspirants	STEM entrants	STEM aspirants	STEM entrants
White	62.00%	61.12%	63.57%	64.11%	67.85%	74.50%
Asian	5.36%	7.11%	8.59%	9.44%	7.58%	9.05%
Black	13.14%	10.41%	13.66%	14.57%	12.02%	7.14%
Hispanic	15.72%	14.60%	9.43%	7.03%	5.76%	5.76%
American Indian	0.80%	1.66%	0.60%	0.25%	0.71%	0.24%
Multi-Racial	2.98%	5.09%	4.16%	4.61%	6.08%	3.30%
URM[3] (total)	**32.64%**	**31.77%**	**27.85%**	**26.46%**	**24.57%**	**16.45%**
TOTAL	100%	100%	100%	100%	100%	100%

Note: Percentages shown are within column. Analyses were based on ELS:2002 data.

1 ELS:2002 respondents were asked in the second follow-up survey to identify the one field of study they would most likely pursue when first enrolled at the community college they attended. Fourteen fields of study were included in the response set, plus "other" and "don't know/undecided." Students who were considered STEM aspirants included those choosing "engineering or engineering technology," "computer or information sciences," or "natural sciences or mathematics."

2 Students' choice of major fields was also measured in 2006. Students were categorized as being STEM entrants if they declared a major in any of the following: "agriculture/natural resources/related," "biological and biomedical sciences," "computer/info sciences/support tech," "engineering technologies/technicians," "mathematics and statistics," "mechanical/repair technologies/techs," "physical sciences," or "science technologies/technicians." It should be noted that in ELS:2002, the response categories to the survey item used in this analysis to measure STEM aspirations and those corresponding the item measuring choice of majors are not the same, with the STEM aspirations measure including much broader categories.

3 URM = Underrepresented racial and ethnic minorities

Although recent high school graduates of 2004 were more likely to aspire to and enroll in STEM disciplines if they began at a four-year institution instead of at a community college, Table 1.1 reveals that the community college groups—both STEM aspirants and STEM entrants—had a distinctively larger share of racial and ethnic minority students who are traditionally underrepresented in the STEM areas of postsecondary study. While we know that community colleges tend to enroll a disproportionately large number of underrepresented racial minorities to begin with, data show that this enrollment pattern also holds true in the STEM areas, justifying the many recent policy propositions and arguments that community colleges represent a unique opportunity in increasing the racial diversity in STEM (Burke & Mattis, 2007; Laanan, 2003; Starobin & Laanan, 2010; U.S. Department of Labor, 2007).

It is noteworthy that over 30% of the STEM-aspiring high school graduates of 2004 enrolled in community colleges by 2006 were underrepresented racial and ethnic minority students. Needless to say, for potential community college bound students who wish to enter STEM fields, cultivating and sustaining their interest would represent an essential first step toward a viable STEM pathway, especially given that academic- and career-related interests are the foundations for future educational and career development.

Prior research has suggested that learning experiences in high school math and science influence the development of STEM interest and have a carry-over effect on actual STEM enrollment in postsecondary education (Wang, 2012). Empirical work has also indicated that high school math and science learning is predictive of future STEM persistence and attainment among students (Crisp, Nora, & Taggart, 2009). However, most of the research in this vein focuses on four-year enrollees, and community college students in STEM have received minimal empirical attention. Therefore, it is essential to explore how the traditional measures of high school learning in math and science, such as number of courses taken and test scores, influence STEM-aspiring students seeking entry into postsecondary education through a community college, in order to advance effective and feasible recommendations for policy and practice.

In addition, existing research has shown that racial and ethnic backgrounds largely impact how high school math and science learning is related to STEM aspirations. In particular, Wang (2012) found that given the same levels of math achievement and courses in math and science, African Americans and Hispanics experience the least gain in their STEM interest compared to their White and Asian counterparts. This difference indicates that the racial/ethnic disparity in STEM participation and success may not simply be accounted for by the historically low levels of academic preparation (Cole & Espinoza, 2008; Perna et al., 2009) in math and science among disadvantaged racial minority students. For community colleges receiving STEM-aspiring underrepresented minority students, it is important to understand how math and science learning better prepare these interested students to enter STEM and what additional factors contribute to their STEM-related aspirations and choices. The issue of increasing participation of underrepresented racial and ethnic minorities in STEM must be tackled by taking into account the possible heterogeneous impacts of various policies and practices so as to more effectively target these students in growing the diversity of the STEM pipeline.

Additional research and practical efforts must be expanded on improving the effectiveness of math and science courses at both high school and postsecondary levels in promoting and sustaining STEM interest of underrepresented racial and ethnic minority students heading toward community colleges. It is important to develop a seamless alignment between secondary and postsecondary math and science offerings and assessment. However, considering the complexity associated with community college entrants' academic and career intentions, expectations, and goals (Bailey, Jenkins, & Leinbach, 2005; Cohen & Brawer, 2008; Laanan, 2003), the developmental process of community college entrants' STEM interest and choice may be more intricate compared to students beginning at four-year institutions. Therefore, future research is warranted to further explore how STEM-related interest and choice behaviors develop among racial and ethnic minority students entering community colleges.

All Beginning Community College Students in STEM: Initial Choice of STEM Major and STEM Transfer

Community College Students in STEM Fields

While ELS:2002 focused on a particular cohort of high school graduates, the 2004/09 Beginning Postsecondary Students Longitudinal Study (BPS:04/09) followed all 2003–2004 beginning postsecondary students' enrollment for over 6 years and thus offers a fuller picture of underrepresented racial and ethnic minority students attending various types of institutions, including community colleges. BPS:04/09 followed a nationally representative sample of students who began postsecondary education for the first time in the 2003–2004 academic year. Respondents were interviewed in their first, third, and sixth year of college. As a supplement to BPS:04/09, the Postsecondary Education Transcript Study (PETS:09) collected transcript data from all the postsecondary institutions attended by the BPS respondents over a 6-year period. Together, BPS:04/09 and PETS:09 provide rich information on enrollment, major fields of study, persistence, and degree attainment from 2003–2004 to 2008–2009, thus representing an ideal data source for studying the evolving role of community colleges in recent years in relation to STEM fields. Approximately 5,550 of all BPS panel respondents began at a public two-year institution, 4,640 at a public four-year institution, and 3,680 at a private four-year institution.

Table 1.2 is a general racial and ethnic profile of beginning postsecondary students in STEM fields when they were first enrolled in the 2003-2004 academic year. Once again, the percentage of underrepresented racial and ethnic minority students relative to White and Asian students is larger at community colleges than those at four-year public or private institutions. Focusing on all beginning postsecondary students, including both traditional and nontraditional age first-time freshmen, data from BPS:04/09 further highlights the important role of community colleges in serving a substantial body of underrepresented racial and ethnic minority students in STEM fields of study.

Often times, these underrepresented students also lack various types of capital necessary for the pursuit of a postsecondary education: academic, social, and economic (Bryant, 2001; Hagedorn & Purnamasari, 2012; Packard, Gagnon, LaBelle, Jeffers, & Lynn, 2011; Tsapogas, 2004). Moreover, given the relatively high level of academic requirements, notably in math, those STEM-aspiring underrepresented minority students may completely lose hope in pursuing a STEM major if they were to begin at a four-year institution. With the open admissions criteria, community colleges have made education accessible to a population that would not have had the opportunity to go to college (Bryant,

TABLE 1.2 Percentage of Beginning Postsecondary Students of 2003–2004 Whose Major During First Year of Enrollment Was in STEM[1]

Race/Ethnicity	Community College Major in STEM	Four-Year Public Major in STEM	Four-Year Private Major in STEM
White	64.26%	66.91%	63.45%
Asian	6.69%	8.07%	10.38%
Black	13.44%	10.89%	10.71%
Hispanic	12.10%	9.19%	12.16%
American Indian	0.07%	0.72%	0.23%
Multi-Racial	1.83%	3.07%	1.82%
Pacific Islander	0.02%	0.29%	0.28%
Other	1.59%	0.85%	0.98%
URM[2] (total)	**29.05%**	**25.02%**	**26.17%**
TOTAL	100%	100%	100%

Note: Percentages shown are within column. Analyses were based on BPS:04/09 data.

1 In the BPS:04/09 survey, 32 fields of study were included in the response set, plus "other" and "undeclared or not in a degree program." Students were categorized as being STEM entrants if they declared a major in any of the following: "agriculture/natural resources/related," "biological and biomedical sciences," "computer/info sciences/support tech," "engineering," "engineering technologies/related fields," "mathematics and statistics," "mechanical/repair technologies/techs," "physical sciences," or "science technologies/technicians."
2 URM = Underrepresented racial and ethnic minorities

2001; Hagedorn & DuBray, 2010), and opened doors to underrepresented racial and ethnic minority students who are interested in STEM but may be deterred from initially attending a four-year institution.

On the other hand, serving a larger share of underrepresented students in STEM fields represents challenges for community colleges. Given the compounded barriers previously noted facing these students, they may experience substantial educational obstacles once at community colleges, especially when they do not possess the criteria necessary for admission to a four-year college or university from the start. In particular, despite their initial academic interest in STEM, underrepresented racial and ethnic minority students may still struggle with subjects essential to STEM disciplines such as math and science (Hagedorn & DuBray, 2010).

Of course, community colleges have a long history of offering developmental and remedial courses (Hagedorn & DuBray, 2010). Successful completion of these remedial courses in math (in particular) may help move academically underprepared students toward continuing to pursue their STEM interest. Math also is the most common area in which students are in need of remedial work (Bahr, 2010). Despite the obvious need for successful remediation in order to better prepare underrepresented racial and ethnic minority students for the rigor of STEM disciplines, previous research shows that African Americans and Hispanic students are more likely to be placed in remedial and developmental courses but are less likely to complete them successfully, highlighting the disparate impacts of the remedial curriculum on these students (Aud et al., 2011). Extant knowledge on remediation, particularly in math, suggests that despite the substantial demand for remedial work in the subject, current curricular and instructional practices do not seem to adequately remedy this need. In order to improve the STEM education experience and pathways for underrepresented racial and ethnic minority students in community colleges, more needs to be done to improve the efficacy of remedial math.

Community College STEM Transfer

Among the wide range of missions of community colleges, upward transfer to a baccalaureate-granting institution has always been a prominent one. Therefore, in discussing the role of community colleges in promoting the representation of racial and ethnic minorities in STEM, it is imperative to examine the extent to which these institutions serve the students who begin their postsecondary career at a community college but seek to earn a bachelor's degree in a STEM field. Based on the BPS:04/09 data, nearly 25% of beginning community college students transferred to a four-year institution over the six-year period. Among those who transferred, almost 20% transferred into a STEM major at a four-year college or university. Table 1.3 below provides the racial/ethnic breakdown of those community college STEM transfer students.

TABLE 1.3 Percentage of Beginning Community College Students of 2003–2004 Who Transferred Into a Four-Year STEM Major[1]

Race/Ethnicity	Transfer into STEM at Four-Year	
	by 2006	by 2009
White	69.38%	62.00%
Asian	13.05%	15.92%
Black	5.10%	10.17%
Hispanic	7.45%	7.85%
American Indian	0.25%	0.35%
Multi-Racial	2.89%	1.84%
Pacific Islander	0.00%	0.00%
Other	1.88%	1.87%
URM[2] (total)	**17.57%**	**22.08%**
TOTAL	100%	100%

Note: Percentages shown are within column. Analyses were based on BPS:04/09 data.
1 In BPS:04/09 survey, 32 fields of study were included in the response set, plus "other" and "undeclared or not in a degree program." Students were categorized as being STEM transfers if they declared a primary major or secondary major in any of the following: "agriculture/natural resources/related," "biological and biomedical sciences," "computer/info sciences/support tech," "engineering," "engineering technologies/related fields," "mathematics and statistics," "mechanical/repair technologies/techs," "physical sciences," or "science technologies/technicians."
2 URM = Underrepresented racial and ethnic minorities

Among community college students who transferred into a four-year STEM major by 2006, about 18% are underrepresented racial and ethnic minorities, and this percentage increased to 22% by the year of 2009. Given these numbers, it is undeniable that community colleges are sending a fair share of underrepresented racial and ethnic minority students to four-year STEM majors. However, compared to the fact that almost 30% of all community college STEM students in 2003–2004 are underrepresented racial and ethnic minorities, these numbers still represent room for improvement.

As previously noted, due to the widely acknowledged academic rigor of STEM disciplines (Perna et al., 2009) at four-year institutions, some baccalaureate-aspiring students interested in these fields of study, including members of underrepresented racial and ethnic minorities, may feel deterred from pursuing such majors at a four-year institution right away. In this sense, the possibility of building the groundwork at a community college may represent a viable and alternative path to the four-year baccalaureate in STEM, thus increasing access and persistence in these fields of study among such underrepresented groups. Policy makers at the national level are increasingly interested in the role of community colleges in helping these students move along the STEM pipeline and whether the institutions have been effective in assisting those students realize their educational goals.

It should be noted that despite the increasing national attention that has been paid to the role of community colleges in expanding the STEM pipeline, the transfer pathways in STEM are still insufficient to meet the demands and therefore need to be further expanded upon (Dowd, 2011). Researchers and policy makers have advocated for well-coordinated efforts to improve STEM education in community colleges, promote community college student success in STEM, as well as foster transfer pathways in these disciplines between community colleges and four-year institutions (National Research Council and National Academy of Engineering, 2012). At the institutional level, understanding how community colleges impact outcomes for STEM-aspiring students can help direct potential resources to improve those outcomes. If more underrepresented racial and ethnic minority students who begin at community colleges can successfully transfer to STEM fields at four-year institutions, we will undoubtedly see an increased diversity among STEM bachelor's degree recipients. In general, evidence-based policy interventions are needed to further support STEM-aspiring students to enter, persist in, and graduate from these challenging and vital fields of postsecondary study.

Four-Year STEM Majors Attending Community Colleges

A less discussed but noteworthy role of community colleges in STEM lies in the fact that community colleges offer classes necessary to apply toward a bachelor's degree (Dowd, 2007; Tsapogas, 2004). Community colleges not only provide the training necessary for associate's degrees or other terminal qualifications, but they also offer foundational coursework that can easily be used toward part of a baccalaureate degree. As a matter of fact, about 17% of BPS:04/09 students who began at four-year institutions and who entered into STEM majors had enrolled to take courses at community colleges during the six-year time span. Although the dataset does not contain information in regard to the specific community college courses (STEM or non-STEM courses) these students took, the community college appears to constitute an important part of the holistic educational experience of these STEM majors beginning at four-year colleges and universities, nearly 30% of whom are underrepresented racial and ethnic minorities (see Table 1.4).

As previously discussed, community colleges are growing to accommodate and offer programs that can transition more seamlessly into bachelor's degrees at other institutions. This not only represents an alternative route for aspiring STEM transfers, but also applies to students attending a four-year institution enrolling in courses concurrently at a community college. In cases where four-year STEM majors may need greater flexibility with general course requirements, they can choose to complete them at a community college. The courses can count toward their degree requirements without delaying their progress at the four-year institution.

TABLE 1.4 Percentage of Beginning Four-Year Undergraduates in STEM Majors Who Ever Attended Community Colleges[1]

Race/Ethnicity	Four-Year STEM Majors Ever Enrolled in Community Colleges
White	60.01%
Asian	10.35%
Black	11.73%
Hispanic	11.70%
American Indian	0.44%
Multi-Racial	3.61%
Pacific Islander	0.00%
Other	2.16%
URM[2] (total)	**29.64%**
TOTAL	100%

Note: Percentages shown are within column. Analyses were based on BPS: 04/09 data.
1   This analysis was first restricted to the students whose first postsecondary institutions were public or private four-year institutions. Then, their STEM major status was identified by any declaration of a primary or secondary major in STEM fields of study during the 2004–2009 time period. If students within this group reported that they ever attended a two-year public institution by 2006 and 2009, they were counted as beginning four-year STEM majors ever enrolled in community colleges.
2   URM = Underrepresented racial and ethnic minorities

Promoting Community Colleges in Serving Minority Students in STEM: Bridging Research and Policy

This chapter drew upon data from ELS:2002 and BPS:04/09, two most recent national datasets that deal with postsecondary participation of both recent high school graduates and returning adults. It provided a general profile in terms of proportions of racial and ethnic minority students entering into STEM fields of study at community colleges, community college students transferring into STEM fields at four-year institutions, and beginning four-year students in STEM fields taking courses at community colleges. Overall, community colleges are serving a substantial body of these students with various goals and pathways as related to STEM.

Facing the important task of better serving the large number of racial and ethnic minority students in STEM in the midst of challenging global economic conditions, community college leaders and their partners from high schools and four-year institutions should search reliable evidence to inform choices and decisions. For community colleges' racial and ethnic minority students to transform their existing STEM interest into an academic action and for four-year STEM students to make the most of their community college education experience, appropriate opportunities and conditions must occur. The promising news is that in recent years, community colleges have begun to take action in order to improve the state of STEM education. Numerous programs at community colleges funded by the federal agencies target STEM participation of underrepresented racial and ethnic minority students (Eagan, Sharkness, Hurtado, Mosqueda, & Chang, 2011; Hagedorn & Purnamasari, 2012; Schultz et al., 2011). These promising programs include mentoring (Packard, 2011; Schultz et al., 2011), faculty support and collaboration (Astin & Astin, 1992; Eagan et al., 2011), supplementary training and instruction (Schultz et al., 2011), research experience (Eagan et al., 2011; Packard, 2011; Schultz et al., 2011), and various others.

Despite the myriad measures of intervention, literature is largely limited to reporting which programs have been introduced and less rigorous evaluation has been conducted in terms of the efficacy of these programs. Hence, research has not pointed to many potential paths policy makers and educators can take in order to better meet the needs of the underrepresented students in the critical areas of STEM study. As Karp (2008) noted, there still exists little evidence-based knowledge on what works to promote student success in community colleges, which begs for sound research addressing issues of institutional innovation that must help their students with a seamless educational experience in and through community colleges. This is especially true for the status of STEM-related programs and research pertaining to community colleges and the many racial and ethnic minority students they serve. Therefore, a most critical recommendation is to build a capacity to use data and evidence to improve STEM program quality and student success at the nation's community colleges.

Given the pressing educational concerns facing STEM, community colleges' indispensable role in addressing them, and the need for valid data and evidence to support related decision making, it is pivotal that rigorous research focusing on student success be conducted and that resulting findings be utilized to improve STEM pathways. Work that integrates research and innovation will greatly help improve community colleges' capacity to better educate and prepare underrepresented racial and ethnic minority students enrolling in STEM programs at these colleges. Policy makers and researchers should build synergetic partnerships to systematically examine and document the effectiveness of programs and offerings in STEM areas on student outcomes. Such evaluation efforts should not only focus on the overall analysis of whether a particular program or intervention works or not, but more importantly, attention should be given to specific features of program and student support offerings to understand *how* they work. In addition, a longitudinal approach will allow for examining the common experience of cohorts of students and factors most relevant to their success as they progress in a particular program.

Community colleges are increasingly recognized as playing a major role in providing relevant and rigorous STEM education and training for students, especially for underrepresented racial and ethnic minority students. Building on nationally representative data, this chapter has pointed to many relevant facets pertaining to this issue. With continued inquiry and analysis driven by relevant policy issues, researchers should be able to work with policy makers and educators in developing sound and viable policy and practice recommendations for community colleges to cultivate a more sustainable stream of STEM-aspiring students, especially those of underrepresented groups. Promoting the long-term educational success of aspiring and current ethnic and minority students in STEM at community colleges is central to addressing the broader issue of access and equity in postsecondary STEM education.

References

Anderson, E., & Kim, D. (2006). *Increasing the success of minority students in science and echnology*. Washington, DC: American Council on Education.

Astin, A. W., & Astin, H. S. (1992). *Undergraduate science education: The impact of different college environments on the educational pipeline in the sciences. Final report*. Los Angeles: University of California, Higher Education Research Institute.

Aud, S., Hussar, W., Kena, G., Bianco, K., Frohlich, L., Kemp, J., &Tahan, K. (2011). *The condition of education 2011* (NCES 2011-033). U.S. Department of Education, National Center for Education Statistics. Washington, DC: U.S. Government Printing Office.

Bailey, T., Jenkins, D., & Leinbach, T. (2005). *Graduation rates, student goals, and measuring community college effectiveness* (CCRC Brief No. 28). New York, NY: Columbia University, Community College Research Center.

Bailey, T. R., Leinbach, D. T., Scott, M., Alfonso, M., Kienzl, G. S., & Kennedy, B. (2004). *The characteristics of occupational sub-baccalaureate students entering the new millennium.* New York, NY: Columbia University, Community College Research Center.

Bahr, P. R. (2010). Preparing the underprepared: An analysis of racial disparities in ostsecondary mathematics remediation. *Journal of Higher Education, 81*(2), 209–237.

Bryant, A. (2001). ERIC Review: Community college students: Recent findings and trends. *Community College Review, 29*(3), 77–93.

Burke, R., & Mattis, M. (2007). *Women and minorities in science, technology, engineering and mathematics: Upping the numbers.* North Hampton, MA: Edward Elgar.

Chubin D. E., May, G. S., & Babco, E. L. (2005). Diversifying the engineering workforce. *Journal of Engineering Education, 94*(1), 73–86.

Crisp, G., Nora, A., & Taggart, A. (2009). Student characteristics, pre-college, college, and environmental factors as predictors of majoring in and earning a STEM degree: An analysis of students attending a Hispanic Serving Institution. *American Educational Research Journal, 46*(4), 924–942.

Cohen, A., & Brawer, F. (2008). *The American community college* (5th ed.). San Francisco, CA: Jossey-Bass.

Cole, D., & Espinoza, A. (2008). Examining the academic success of Latino students in science technology engineering and mathematics (STEM) majors. *Journal of College Student Development, 49*(4), 285–300.

Dowd, A. C. (2007). Community colleges as gateways and gatekeepers: Moving beyond the access "saga" toward outcome equity. *Harvard Educational Review, 77*(4), 407–418.

Dowd, A. C. (2011). *Developing supportive STEM community college to four-year college and university transfer ecosystems.* A report prepared for Community Colleges in the Evolving STEM Education Landscape: A Summit, Washington, DC.

Eagan, M. K., Sharkness, J., Hurtado, S., Mosqueda, C., & Chang, M. J. (2011). Engaging undergraduates in science research: Not just about faculty willingness. *Research in Higher Education, 52*(2), 151–177.

Hagedorn, L. S., & DuBray, D. (2010). Math and science success and nonsuccess: Journeys within the community college. *Journal of Women and Minorities in Science and Engineering, 16*(1), 31–50.

Hagedorn, L. S. & Purnamasari, A. V. (2012). A realistic look at STEM and the role of community colleges. *Community College Review, 40*(2), 145–164.

Hurtado, S., Han, J. C., Saenz, V. B., Espinosa, L. L., Cabrera, N. L., & Cerna, O. S. (2007). Predicting transition and adjustment to college: Biomedical and behavioral science aspirants' and minority students' first year of college. *Research in Higher Education, 48*(7), 841–887.

Karp, M. M. (2008). *Towards a community college research agenda: Summary of the National Community College Symposium,* Washington, DC. Retrieved from http://ccrc.tc.columbia.edu/Publication.asp?uid=662

Laanan, F. (2003). Degree aspirations of two-year college students. *Community College Journal of Research and Practice, 27*(6), 495–518.

May, G., & Chubin, D. (2003). A retrospective on undergraduate engineering success for underrepresented minorities. *Journal of Engineering Education, 92*(1), 27–40.

Museus, S. D., Palmer, R. T., Davis, R. J., & Maramba, D. C. (2011). Racial and ethnic minority students' success in STEM education [Special report]. *ASHE Higher Education Report, 36*(6).

National Academies 2005. "Rising Above the Gathering Storm" Committee. (2010). *Rising above the gathering storm: Rapidly approaching Category 5.* Washington, DC: Author.

National Research Council and National Academy of Engineering. (2012). *Community colleges in the evolving STEM education landscape: Summary of a summit.* S. Olson & J. B. Labov, Rapporteurs. Planning Committee on Evolving Relationships and Dynamics Between Two- and Four-Year Colleges, and Universities, Board on Higher Education and Workforce, Division on Policy and Global Affairs; Board on Life Sciences, Division on Earth and Life Studies; Board on Science Education, Division on Behavioral and Social Sciences and Education; Engineering Education Program Office, National Academy of Engineering and Teacher Advisory Council, Division on Behavioral and Social Sciences and Education. Washington, DC: The National Academies Press.

Obama, B. (2009, February 24). Address before joint session of Congress. Retrieved from http://www.whitehouse.gov/the_press_office/Remarks-of-President-Barack-Obama-Address-to-Joint-Session-of-Congress/

Packard, B. W. L. (2011, December). *Effective outreach, recruitment, and mentoring into STEM pathways: Strengthening partnerships with community colleges.* Paper presented at the Community Colleges in the Evolving STEM Education Landscape, Washington, DC.

Packard, B. W. L., Gagnon, J. L., LaBelle, O., Jeffers, K., & Lynn, E. (2011). Women's experiences in the STEM community college transfer pathway. *Women and Minorities in Science and Engineering, 17*(2), 129–147.

Perna, L., Lundy-Wagner, V., Drezner, N. D., Gasman, M., Yoon, S., Bose, E., & Gary, S. (2009). The contribution of HBCUS to the preparation of African American women for stem careers: A case study. *Research in Higher Education, 50*(1), 1–23.

Schultz, P. W., Hernandez, P. R., Woodcock, A., Estrada, M., Chance, R. C., Aguilar, M., & Serpe, R. T. (2011). Patching the pipeline: Reducing educational disparities in the sciences through minority training programs. *Educational Evaluation and Policy Analysis, 33*(1), 95–114.

Smyth, F. L., & McArdle, J. J. (2004). Ethnic and gender differences in science graduation at selective colleges with implications for admission policy and college choice. *Research in Higher Education, 45*(4), 353–381.

Starobin, S., & Laanan, F. (2010). From community college to Ph.D.: Educational pathways in science, technology, engineering, and mathematics. *Journal of Women and Minorities in Science and Engineering, 16*(1), 51–66.

Tsapogas, J. (2004). *The role of community college in the education of recent science and engineering graduates* (NSF 04–315). National Science Foundation: Arlington, VA.

U.S. Department of Labor. (2007). *The STEM workforce challenge: The role of the public workforce system in a national solution for a competitive science, technology, engineering, and mathematics (STEM) workforce.* Washington, DC: U.S. Department of Labor.

Wang, X. (2009). Baccalaureate attainment and college persistence of community college transfer students at four-year institutions. *Research in Higher Education, 50*(6), 570–588.

Wang, X. (2012). *Modeling student choice of STEM fields of study: Testing a conceptual framework of motivation, high school learning, and postsecondary context of support* (Working paper). Madison: Wisconsin Center for the Advancement of Postsecondary Education.

2

THE IMPACT OF STATE POLICY ON COMMUNITY COLLEGE STEM PROGRAMS

Pamela Eddy

Declining college graduation rates, poor performance in mathematics and science, and an increasingly competitive and technology-driven global economy breed a setting for education reform in the United States. This environment pressures leaders at all levels of government to take action with a focus to open the pipeline in STEM disciplines, ultimately with the intention of obtaining more college graduates in these high demand areas. National and state initiatives are calling for more college degrees and workforce training to strengthen America's workforce and its economic prosperity (2011 Higher Education Opportunity Act of Virginia, 2011; Obama, 2009). The American Graduation Initiative (Obama, 2009) targeted a goal of achieving 5 million more college degrees by 2020; likewise, the Lumina Foundation (2010) established its "Big Goal" that targets 60% of Americans having a postsecondary degree or credential by 2025. Reaching these lofty ambitions requires progress within each state. In 2011, Virginia passed its Higher Education Opportunity Act, also known as Top Jobs for the 21st Century (Top Jobs Act or TJ21) that set a statewide target of 100,000 additional degrees by 2025. If all 50 states had similar targets, the goal of 5 million additional degrees by 2025 would be possible.

The Top Jobs Act outlined 10 purposes to guide development and implementation of funding policies and to establish evaluation criteria. In summary, the purposes include:

1. To ensure an educated workforce in Virginia;
2. To take optimal advantage of the demonstrated correlation between higher education and economic growth;
3. To place Virginia among the most highly educated states and countries by conferring approximately 100,000 cumulative additional undergraduate degrees on Virginians between 2011 and 2025;

4. To enhance personal opportunity and earning power for individual Virginians by increasing college degree attainment in the Commonwealth, especially in high-demand, high-income fields such as science, technology, engineering, mathematics (STEM), and health care;
5. To promote university-based research that produces outside investment in Virginia to fuel economic growth;
6. To support the national effort to enhance the security and economic competiveness of the United States of America through increased research and instruction in science, technology, engineering, mathematics, and related fields (STEM);
7. To preserve and enhance the Virginia higher education system's excellence and cost-efficiency through reform-based investment, technology-enhanced instruction, sharing of instructional resources between and among colleges and universities, and expanded community college transfer options leading to bachelor's degree completion;
8. To realize the potential for enhanced benefits from the Restructured Higher Education Financial and Administrative Operations Act of 2005 (§ 23-38.88 et seq.);
9. To establish a higher education funding framework and policy that promotes stable, predictable, equitable, and adequate funding, provides need-based financial aid for low-income and middle-income students and families, and relieves the upward pressure on tuition associated with loss of state support;
10. To recognize the unique mission and contributions of each institution of higher education in the Commonwealth (§ 23-38.87:10).

For the purposes of this chapter, this research study focused on the intersection of the points outlined in TJ21 that refer to increasing the number of graduates in STEM fields (4 and 6 above) and the leveraging of community college pathways to enhance college completion (7, 9, and 10 above). The legislation emphasizes the need for developing economic efficiencies and promoting partnerships and collaborations among public institutions and private partners (§ 23-38.87:19). The research questions at the heart of this study included: How does state higher education policy influence planning efforts at community colleges? What steps are Virginia community colleges taking to help support educational efforts in STEM?

The literature review presents background on the policy implementation process, central steps required for strategic planning and change, and current research on STEM pathways in community colleges. Next, a review of the data collection methods employed for this study outlines the research design process. A case study portrait sets the context for the findings from the study. Finally, the chapter concludes with a discussion of the data and a summary of recommendations for practice.

Literature Review

The strong link between demand to fill STEM jobs and the emphasis on economic development in cities and states focuses state leaders' and policy makers' attention on these issues. The National Governors Association (NGA) acknowledged the critical nature of student preparation in STEM on future prosperity for the country, noting:

> Governors are in a unique position to advance comprehensive STEM education policy agendas aligned with workforce expectations that will ultimately aid state economic growth. Governors can elevate the urgency and build the political will to advance STEM education and use budgetary and policy levers to make meaningful changes across education systems.
>
> *(NGA, 2011, ¶ 3)*

Scaffolding and coordination within the educational pipeline becomes central to state policies to support STEM education. The Department of Labor reported that "of the 20 fastest growing occupations, half are in the associate degree or higher category" (Bureau of Labor Statistics, 2012, p. 13). Thus, the pathways through community colleges become increasingly important for efforts to expand graduates in STEM fields. Minor (2012) underscored the importance of community colleges in this pipeline, in particular for minority students as the majority of Hispanics and African Americans holding a bachelor's or master's degree in the STEM field have attended a community college. Using state policy as a lever for change influences the approaches community colleges pursue to support STEM education.

Policy Implementation

The research on the policy process is robust and includes consideration for the development, adoption, implementation, and evaluation phases of the policy process (Fowler, 2009). Case studies of policy implementation targeting STEM support and completions, however, are limited (Johnson, 2012). The scant research on STEM policy reinforces the challenges facing policy implementers that contribute to the overall success or failure of the implementation (e.g., lack of motivation to implement at the grassroots' level, scarcity of resources, or mismatch of policy intentions and community needs) (Breiner, Harkness, Johnson, & Koehler, 2012; Fowler, 2009).

Fowler (2009) described the growing field of implementation research by dividing it into three "generations" (p. 272) according to eras that focused on different emphases regarding the effectiveness of implementation strategies. The first generation of implementation research revealed the difficulty of implementing policy. According to Fowler (2009), policy implementers, specifically intermediaries who have the responsibility to actually implement policy, often lack the understanding, knowledge and skills, and resources necessary to implement the policy. The second-generation research reemphasized the challenges with policy implementation, but also focused on the design of the policies, identifying factors that were present when policies were successfully implemented. The third, and current, generation considers more carefully implementers as learners and is based on the concept of schemas, the natural tendency to rely upon previous experiences to relate to new information (Harris, 1994; Weick, 1995). Furthermore, the third-generation era of research on implementation stresses the importance of a strong social infrastructure, as an arena of idea generating and sharing network, in policy implementation (Fowler, 2009).

Through the lessons learned from the three generations of implementation research, today, we are able to identify patterns and commonalities within successful implementation processes. Fowler (2009) offered a "how to" for implementing policy, in which great significance is placed upon the mobilization process, including the steps of policy adoption, planning, and the gathering of resources. The mobilization phase typically lasts from 14 to 17 months (Huberman & Miles, 1984). Thus, in 2012, Virginia was still in the middle of this process with respect to the Top Jobs Act. At the outset of a new policy, three guiding questions prevail: (a) Is there good reason to adopt policy? (b) Is the policy appropriate for the institution and its serving region? (c) Is there sufficient support for the policy among key stakeholders? (Fowler, 2009, pp. 256-258). These questions help guide the analysis of the data for this study. Influencing the tension between planning too much and planning too little is gathering the appropriate and adequate resources for successful policy implementation. For instance, Henderson, Beach, and Finkelstein (2011) conducted an analytic review of the literature regarding changes to instructional practices in undergraduate STEM courses and found that top-down institutional mandates for change were ineffective, whereas effective change strategies involved alignment of values and beliefs about the policy and teaching strategies.

Strategic Planning and Change Theory

A number of planning and change models exist. Keller's seminal 1983 book on academic strategy in higher education ushered in the use of strategic planning and management within colleges and universities. This time period represented the mark of declining state support for higher education and a shift in public sentiment of higher education from a public good to more of a private good (Bowen, 1977). Planning became increasingly crucial for institutional survival. Leaders sought clear and concise outlines for planning efforts. Rowley, Lujan, and Dolence (1997) created a 10-step cyclic model that involved establishment of key performance indicators (KPIs), environmental scanning, brainstorming and formulation of ideas, development of implementation strategies, and continuous evaluation. Yet, strategic planning is not always at the center of policy development as ideologies, power, and the timing associated with defining issues often influence issue definition and agenda setting (Fowler, 2009). A focus on change, however, is a commonality for policy and planning efforts.

Fullan (2002, 2006) provides a framework for change specifically targeting education. His research on change theory challenges us to consider "what 'theories of action' really get results in educational reform"? (2006, p. 3). Fullan's (2006) model contains seven core premises: focus on motivation, capacity building with a focus on results, learning in context, changing context, reflective action, tri-level engagement, and persistence and flexibility in staying the course. Central to this framework are motivation and engagement because the remaining five premises depend upon the synergy of the desire to change and the willingness to participate in action. The implementation phase of policy involves change and depends on the implementers' motivation for enacting the policy and their capacity or ability to influence how the policy applies in practice (Fowler, 2009). The discourse presented in the early policy process stages of policy development and adoption lays the foundation upon which the later stages of the policy process, specifically implementation and evaluation, are built. According to Fowler (2009), it is in these early stages that "a misguided policy adoption process undermines the entire implementation" (p. 285). Fullan (2006) would argue that "if one's theory of action does not motivate people to put in the effort—individually or collectively—that is necessary to get results, improvement is not possible" (p. 8). Many policies do not result in change due to resistance on-the-ground.

Johnson (2012) utilized Fullan's (2006) theoretical framework of change to study the implementation of a state STEM policy. The study reported several successful strategies including the use of early conversations about shared vision, the development of a strategic plan to guide change, the creation of a structure supporting accountability, and the celebration of small wins. Despite the challenges of implementing the state policy due to funding and timeline and competing agendas of the stakeholders, Johnson (2012) offered hope that successful policy implementation is possible, particularly when a dedicated leader builds the case for change, manages and communicates the message of change with constituents, charts the course of action, and keeps all involved parties on track towards the established goals.

STEM in Community Colleges

Recent research on STEM in community colleges focuses primarily on the experiences of students in general (Hagedorn & Purnamasari, 2012; Heidel et al., 2011; Lenaburg, Aguirre, Goodchild, & Kuhn, 2012) and students of color specifically (Malcom, 2010; Reyes, 2011) versus how state policy regarding

STEM is implemented within the sector. Knowing more about the student experience in the classroom can help inform policy formation and identify and leverage best practices that support student success. Likewise, vocational education and contract training provides a forum to learn more about what helps students in these learning environments complete their programs. A study conducted by the Business Higher Education Forum (BHEF, 2010) sought to create a framework to enhance stakeholder understanding of the community college in workforce development and degree attainment. This research offered two broad frameworks. The first, a regional model, focused on the interactions of the community college, government, and workforce to increase the number of graduates. The second, a sectoral model, focused specifically on a single labor market or profession. Both of these frameworks center on the role of the community college as an actor in discovering solutions to workforce training and education. Indeed, community colleges are increasingly recognized as a potent lever for economic development (Obama, 2009).

The National Governor's Association (NGA; 2011) published an issues brief offering several insights of the advantages of utilizing community colleges to meet the demands of today's workforce. The brief presented arguments to support expanding STEM education and STEM-related workforce skill development, including the need to expand regional STEM-skill needs of the industry, providing incentives for STEM course completion at the institutional and individual student levels, supporting more effective mathematics remediation, and creating requirements for transferable community college STEM credits and credentials (NGA, 2011). With 90% of the U.S. population living within 25 miles of a community college and the traditionally low-cost tuition rates at these colleges (Cohen & Brawer, 2008), community colleges are an attractive and convenient option for bolstering a STEM-skilled workforce. Their return on investment, estimated at 16%, further contributes to the reliability of community colleges to fill in the gaps and in order to do so, efforts will need to be made to address the "policy gaps" (NGA, 2011, p. 1).

Virginia Case Background

Research highlights how state-based policy initiatives are a response to the convergence of federal science and economic policy, with a focus on high tech economic activity across the states (Douglass, 2007). A survey of 22 Virginia Community College workforce development leaders found that healthcare skill training was the highest demanded need in their regions, followed by STEM training in general (Landon, 2009). Central to building technology training capacity are higher education and business and industry collaboration. The Virginia Business Higher Education Council created a 2020 vision to prepare Virginia's workforce for the top jobs of the future initiating the Grow by Degrees Coalition (see http://growbydegrees.org/). In 2009, Bob McDonnell, then

Attorney General, ran on a campaign platform to raise the bar for educational attainment "to bring strong and sustainable economic opportunity and expansion" to the Commonwealth (McDonnell, 2009, ¶ 6). Fulfilling this campaign promise, Governor McDonnell created the Higher Education Commission that helped craft the 2011 legislation for what is now the Top Jobs Act.

When the Top Jobs legislation was signed in 2011, early enrollment projections predicted that Virginia institutions would add 6,000 new seats for in-state students. Of this total, 4,000 of the 6,000 expected students enrolled in a Virginia Community College (DuBois, 2011), which was more than anticipated. The fact that two-thirds of new enrollment occurred at the community college level underscores the important role two-year colleges have in preparing graduates for work demands. The State Higher Education Executive Officers (SHEEO) organization reported on degree completion by program areas for all states (2011). For STEM degree programs in Virginia, community colleges show a completion rate of 26.3% within the sector, which is 17.25% higher than the national average and only slightly below the completion rates for the highest level research universities in the state (Research, high activity—16.00%; Research, very high activity—33.48%). As such, community colleges play a significant role in STEM education in the state of Virginia.

Projections indicate that the Virginia will have the second highest proportion of STEM jobs as a fraction of job openings through 2018 (8.2%); only the District of Columbia has a higher rate (Carnevale, Smith, & Melton, 2011). However, the state has a lower anticipated rate of STEM jobs for those with an associate's degree (8.3% compared to the high in North Dakota of 23.8%). A full 93% of STEM jobs in Virginia will require a postsecondary degree (Carnevale, Smith, Stone, et al., 2011). The bulk of these jobs will require a bachelor's degree or higher, and community colleges provide a prime gateway into these degree programs given their convenience and cost. An anticipated 30% increase in STEM jobs is predicted in Virginia, which is 13% higher than the national norm.

The state of Virginia recognized these facts regarding demands for graduates as the Top Jobs Act reflects with its focus on STEM education and the educational pipeline. Further, Northern Virginia Community College (NOVA; 2012) initiated a program, SySTEMic Solutions, to expand the STEM pipeline and provide a pathway for students from area high schools to NOVA, to George Mason University, and into the workforce. Counselors are embedded into regional high schools and work to identify students for the program and support them during the application process. Once at NOVA, retention counselors continue to support these students as they transition to college and ultimately transfer to a four-year college or into the workforce. These support mechanism recognize the fact that many of the participating students are first-generation college students who are often low-income and from minority groups.

Project Background

The methods for this research included a review of the Virginia Higher Education Opportunity Act of 2011 to determine the influence of the legislation on STEM programming and the anticipated role for community colleges. As noted above, several factors were included in the TJ21 that target both of these areas. Next, a review of the strategic plan (Achieve 2015) for the Virginia Community College System (VCCS) occurred. The features of the plan are highlighted in the findings section. As well, the strategic plans of the 23 community colleges within the state were analyzed to determine links with the overarching state planning document, in particular noting plans for STEM programming and implementation plans for the campuses. Finally, interviews were done with key stakeholders in the state to understand their perception of the implementation phase of TJ21 and to note how change was occurring on the state's community college campuses.

Findings

A summary of the planning efforts at the Virginia Community College System provides a background to understand better how the new Top Jobs Act policy, in particular the focus on STEM programming, has begun to be implemented in the community college system and on individual campuses. The relatively recent passage of the legislation and the typical timeline for policy implementation places this snapshot in the early stages of the process. This initial investigation provides a benchmark for subsequent evaluation in the future. This study found three main findings. First, for the VCCS was already engaged in updating their strategic plan, thus the implementation of the goals outlined in the TJ21 state policy had to be incorporated into a planning process already underway. Second, the change obtained so far in the implementation of the policy has been incremental and first-order change (Bartunek & Moch, 1987). Finally, the state policy implies a coordination of efforts along the educational pipeline that occurs in a limited fashion in reality.

Context of the VCCS

The Virginia Community College System was established in 1966 by the General Assembly to better serve the needs of local communities through the provision of educational and vocational training opportunities. The mission of Virginia's Community Colleges is to give everyone the opportunity to learn and develop the right skills so lives and communities are strengthened (Virginia's Community Colleges, 2012). There are 23 community colleges in the state operating on 40 campuses. The majority of the colleges are serving rural areas (17), with the remaining campuses located in suburban areas (5) or urban cities (1). The campuses range in size from small (enrollment of 1,052) to large (enrollment of 48,996). Three of every five undergraduate students in the state attend one of Virginia's community colleges (VCCS fast facts, 2012). Virginia also has one junior college, Richard Bland College (RBC) that is not in the VCCS. Instead, RBC is a branch campus of the College of William and Mary. As such, RBC has its own unique planning efforts underway that are not coordinated in the VCCS Achieve 2015 matrix; RBC was not included in this study.

The chancellor of the system, Glenn DuBois, utilizes broad-based strategic planning. In 2003, he oversaw a planning effort titled Dateline 2009 in which targets were established for student recruitment, graduation, and transfer rates. This initial 6-year plan was updated in 2009 and the new plan is titled Achieve 2015. Five key areas were identified for this plan: Access, Affordability, Student Success, Workforce, and Resources (see http://www.vccs.edu/WhoWeAre/Achieve2015.aspx). The objectives of Achieve 2015 are in line with national and state priorities, supporting the preparation of a strong workforce for the jobs of the future. The student-first agenda charts the course for Virginia's Community Colleges in an effort to strengthen lives, communities, and ultimately the Commonwealth's economy. Each of the 23 colleges in the state has individual campus plans that align, or not, with the overarching state plan. These individualized plans allow the campus to capture the unique context of their location and to capitalize on their areas of expertise, as well as acknowledge the differences in size among the campuses. Individual campus presidents are evaluated using a set of president's goals that are linked to the overarching system plan.

The Virginia Community College System was able to leverage the language already established in their aggressive strategic plan, Achieve 2015, to outline their plan of action for meeting the goals of TJ21. Aligning with the objectives of national and state initiatives to confer more degrees, particularly in the area of STEM degrees, the VCCS is targeting improvements to remedial instruction and student success. The VCCS is implementing a redesign of developmental education and other reengineering efforts to help support the policy goals of increased numbers of graduates.

Hopping on a Moving Train

The State Council of Higher Education for Virginia (SCHEV) requires 6-year plans from colleges. As a result, the leaders of the community colleges already had in place a framework for planning and change when TJ21 was passed. The colleges could look at their existing plans and determine how they could align plans underway with objectives of the state policy.

The goals of Achieve 2015 are further distilled into strategies, referred to as the chancellors' goals. These goals are evaluated each year and adjustments are made to the language to ensure that VCCS and its campuses are addressing the most current issues with its progress towards Achieve 2015. In particular, under the Access goal, one of the strategies focuses on educational programs and states that the objective is to "annually develop 10 new academic programs (degree, certificate, or career studies certificate) that respond to emerging, critical workforce needs, particularly in *STEM-related areas* (science, technology, engineering, and mathematics)" (D. VanCleave, personal communication, April 3, 2012, emphasis added). In the first year of Achieve 2015, the 2010–2011 chancellor's goals did not include the emphasis on developing STEM-related academic programs. It is inferred that this STEM reference was included later in support of the adoption of TJ21, which took place in April, 2011, and only 1 month before the adoption of the 2011–2012 chancellor's goals in May, 2011.

The review of the individual campus strategic planning documents highlighted a range of alignment with the overarching system goals outlined in Achieve 2015. Eleven of the campuses had alignment with the central plan and included specific references to some of the chancellor's goals. Of this subset, one rural-serving, medium-sized college developed a specific plan for STEM that included several goals and sub-goals to guide the college's progress. Another 11 of the campuses had mixed alignment with the system goals. In these instances, the colleges had a framework in place similar to the system plan, but the language used to communicate the goals did not mirror Achieve 2015. Only one college (suburban, multi-campus) did not have a plan linked from the VCCS website and there was no mention of Achieve 2015 on the college's webpage. As well, the language used to communicate the college's goals did not mirror Achieve 2015. Table 2.1 provides a summary of the alignment of the overarching plan and the individual college's plans.

The majority of rural campuses in the state have plans that align with the overarching state planning document (59%), whereas the bulk of suburban colleges have mixed alignment (80%). Further analysis of the individual campus plans found approximately half (48%) of the campuses with specific references to STEM. The plans referred specifically to the goal of increasing access. This focus is appropriate given the references in the Top Jobs Act to increasing graduates in STEM fields. Planning references on individual campuses targeted creation of new academic programs, with several referencing addressing specific community based and critical workforce needs to determine what programs to develop. One campus identified the goal to increase enrollment of at-risk and underserved students, particularly mentioning women and minorities in

TABLE 2.1 Alignment of College Plans with State Plan

Alignment	Number of Colleges	College Classification
Aligns	11 (48%)	10 rural; 1 urban
Mixed	11 (48%)	7 rural; 4 suburban
Not Align	1 (4%)	1 suburban

STEM programs. Two of the campus plans outlined the creation of a new Academic Technology Center or building a new center. One campus took the perspective that strategic faculty hiring was required to increase the number of STEM graduates, pointing out that they sought to obtain a 40:60 ratio of tenured faculty to adjunct faculty members. Since 2000, a total of 29 new STEM programs have been created within the VCCS (SCHEV, 2012). Fourteen of the colleges created one or more programs in this timeframe. New programs were created on 11 of the 17 rural campuses and three of the five suburban campuses.

Change or Wordsmithing?

The objective to increase graduates in the state and to bolster programming for STEM degree majors resulted in the articulation of particular goals and objectives for the state's colleges. As noted above, the planning efforts of the VCCS readily accommodated inclusion of the goal to increase graduations in STEM areas. Bartunek and Moch (1987) argued that first-order change is incremental in nature and occurs within existing frameworks, whereas second-order change involves deeper changes to the existing frameworks. The first year of implementation of the TJ21 policy involved first-order changes. The planning process underway accommodated the changes required in the policy, but the beginning stages of second-order change may be emerging as the VCCS focuses on issues around developmental education. The role of development education is discussed more fully in the following section on the alignment of the educational pipeline.

Given the first-order change underway, it is important to investigate the role language plays in how the state policy is being implemented. How leaders frame change for campus members matters in how these stakeholder groups make meaning of the changes underway (Eddy, 2003; Neumann, 1995). The placement of the focus on STEM within the campus website and strategic plan begins to highlight the role it has on campus. As noted above, less than half of the colleges (48%) had specific references to STEM in their planning documents. Both the chancellor of VCCS and the vice chancellor of Academic Services & Research were interviewed to determine how TJ21 requirements were being integrated into the colleges in the state. Both central office leaders noted how New River Community College (NRCC) represented a leading force in developing STEM programs ahead of the mandate from the state. The current president of NRCC, Dr. Jack Lewis, added, "STEM is certainly not new to New River Community College." The college operates a facility at the mall that offers a state-of-the-art facility. According to the college's website, "The mall location features 14 classrooms, over 200 computers for student use, a science lab, two auditoriums, testing and conference rooms and office spaces" (Lewis, n.d.). Symbolically, having STEM programs concentrated in a location with high visibility sends a message to the community, students, and employers.

According to Chancellor DuBois, some of the changes within the system since the passage of TJ21 have been a in the reward structure. He noted, "Modifications were made to the funding formula to include funds for enrollment growth, and performance funding for STEM-related initiatives. STEM is what is hot and the capital request for buildings and equipment were made in support of STEM" (personal interview, April 19, 2012). Funding has been used as an inducement for change. As President Lewis of NRCC stated, "The community colleges are in a unique position of being the shining star in bearing up and responding to this legislation because we have two-thirds of all undergraduates in Virginia" (n.d.). Recognition is apparent for the need to align with the state's priorities.

The rhetoric around the state goals and ambitions for TJ21 are apparent in the language of the VCCS Achieve 2015. Yet, as the vice chancellor summarized, "The performance measures that are to be developed and the linage to STEM will serve as a testing ground for how serious we are about STEM— which will drive action" (personal interview, Susan Wood, Vice Chancellor, May 2, 2012). The ability of TJ21 to have a lasting and significant impact will be dictated by continued state focus and support in this arena. Colleges will respond when they are rewarded for their behaviors. TJ21 did not propose a particular systematic change or alteration to the underlying framework of higher education.

How leaders frame change on campus also impacts how campus members will react (Fairhurst & Sarr, 1996; Weick, 1995). As President Lewis reflected, "I'm always amazed at how expectations lead to outcomes" (n.d.). One of the ways changes concerning STEM are framed is via the college's and system's strategic planning documents, location of references to STEM on the website, institution of new programming targeting STEM areas, creation of reward structures to support STEM activities, and removal of barriers for students to pursue STEM majors. As Lewis offered, "We put emphasis on the success story at award ceremonies. For example, recently a former student spoke with a stirring speech that was so honest with the students about what it is to be a nurse … Role models are brought in for every pinning ceremony and graduation to raise the expectations and the bar" (n.d.). The fact that only half of the colleges, however, have specific references to STEM highlights the early stages of implementation of the policy. Because the presidents are evaluated on the ways in which they reach the goals set by the chancellor, leverage for further change is possible. A true test of implementation will be in assessing outcomes of heightened STEM programming and focus and STEM graduation rates over the next five years.

Coordination in the Educational Pipeline

The focus of TJ21 on creating pathways through the educational system calls attention to community colleges that serve as bridging institutions between high schools and four-year universities. An historic feature of community colleges is articulation agreements with four-year colleges. As the chancellor noted, "Transfer relationships are good, so the next step may be to create more 3 + 1 arrangements" (personal interview, Glenn DuBois, April 19, 2012). Already in place in Virginia were transfer grants. Beginning in 2007, eligible transfer students could obtain $1,000 if they graduated with an associate's degree, met academic requirements, and were accepted in one of the state's four-year universities the fall after their community college graduation (Two-Year College Transfer Grant Program, 2007, § 23-38.10:9). An additional $1,000 is available for students who pursue a STEM degree. As the vice chancellor for Academic Services and Research noted, "Coupled with the guaranteed admission agreements, STEM could be furthered for those students who want to transfer, creating a powerful tool for families to get access to selective 4-years and financial support by coming to a 2-year college first" (personal interview, Susan Wood, Vice Chancellor, May 2, 2012).

Influencing student success at the community college is college readiness. A two-prong approach is underway to address issues of deficiency. First, more communication is occurring about college expectations in high schools. Second, VCCS is extensively revising their developmental math and English programs. The first of these changes to delivery of developmental programming were implemented in spring 2012 for math, whereas the initiatives for developmental English will begin in spring 2013. A *Chronicle* article highlighted the changes to programming in Virginia, summarizing: "Its (VCCS) colleges will soon replace their semester-long developmental-math courses with nine units, which can be taken as one-credit classes or Web-based lessons with variable credit hours that allow students to complete more than one unit in a self-paced computer lab and classroom" (Gonzalez, 2011, ¶23). The changes were implemented without a formal pilot program and represent a clear break from past practice. This type of fundamental change is more in line with second-order or deep organizational change as it represents questioning of the assumptions of the existing system and using data to change operations (Bartunek & Moch, 1987).

As the vice chancellor reflected, "Another barrier to increasing STEM graduates is a cultural barrier to innovation; thinking that the way we have always done it is the way we should do it" (personal interview, Susan Wood, Vice Chancellor, May 2, 2012). The increased use of data to assess operations and the efforts supporting strategic planning begin to address this historical area of challenge in change initiatives. At NRCC, a partnership between the college and its neighboring four-year university helps support STEM training programs. Of note, NRCC's nursing program enrolls a number of students who already have bachelor's degrees but are seeking specific training. Townsend (1999) referred to this trend as reverse transfer, which is becoming more common for technical programs and career re-tooling.

On the other side of the educational spectrum are dual-enrollment programs. As noted above, Northern Virginia Community College created the SySTEMic

Solutions model, a recognized program to open up the STEM pipeline from high schools to the community college and work or transfer. As well, another program already underway is the placement of career coaches in nearly half of the high schools state-wide. According to Dubois, "There are approximately 150 career coaches in local high schools and they talk about STEM." DuBois added, "Enrollment increased by 8% when career coaches were placed in the high schools; they are a game changer" (personal interview, Glenn DuBois, April 19, 2012). This type of bridge building program allows for a more seamless flow of students between K-12 schools and the community college.

Discussion and Conclusion

The analysis on the implementation process of the Virginia state policy (TJ21) to expand the number of graduates in the state and to enhance a focus on STEM highlights a number of conclusions. The work already underway by the VCCS with both their strategic planning and their reengineering efforts provided a context prime for implementing the changes outlined in the higher education legislation as mechanisms were already in place. The legislation resulted in slight nuances made to existing plans and policies versus a wholesale change effort. What the policy allowed was a way for leaders of community colleges to frame their efforts on campus to emphasize the activities underway to enhance STEM programming and training. The strategic plans of the college created a template for some of this effort, as did the college's websites, technical centers, and STEM-oriented events. Leaders help campus members and college stakeholders to make sense of changes on campus. Involvement in the campus strategic planning process provides one venue for this sensemaking to occur for campus members (Weick, 1995).

Pointedly, the chancellor of the VCCS serves in the role of "*sense-giver*" (Thayer, 1988, p. 250, italics in the original). The messages sent out of the chancellor's office regarding the importance of the Achieve 2015 planning document are clear and numerous. The prominence of the plan on the system website and the references to the planning document on individual campuses in the system highlight how the chancellor has framed the importance of planning to heighten student success. The passage of TJ21 resulted in the VCCS evaluating the Achieve 2015 planning documents to accommodate for the requests of the policy, namely the focus on STEM. The existing plans already targeted increasing the number of graduates from the state's community colleges.

A mechanism to help create and reinforce the focus on STEM in community colleges is the use of an institutional saga (Clark, 1972). As noted in the findings, President Lewis of NRCC used graduation events and pinning ceremonies to provide student testimonials about the importance of their technical training to their ultimate careers. Moreover, the presence of the technical center in the mall became part of the saga of the college's approach to innovation.

The SySTEMic Solutions program at NOVA reinforces a saga revolving around technology too. Because of the infusion of technology into the area high schools as a result of building a bridge with dual enrollment through the program, the saga is reinforced in the community.

Yet, not all the colleges in the VCCS have this level of prominence in promoting STEM. On the one hand, the implementation phase of TJ21 is still in its infancy and the focus might be in early stages on some of the campuses. On the other hand, the regional needs of the colleges may differ such that how STEM programming needs are interpreted or valued may depend on location. For instance, small rural schools may find that their limited resources result in programming that looks markedly different than their larger, suburban counterparts. As well, the needs of employers in the region may differ. The projected increase of need for STEM professionals in the state may mask some of the regional differences. Investigation into deconstruction of anticipated demand should occur for more targeting programming to result.

Barriers to full implementation of the TJ21 legislation revolve around funding, access, and college readiness. Current funding is rewarding colleges that focus on STEM programming, but if this program funding is eliminated, the continued support of STEM might be jeopardized. McDonnell and Elmore (1987) identified four main frameworks employed by policy makers to obtain change. These include mandates, inducements, capacity-building, and system-changing. The TJ21 Act falls within the area of inducements and capacity-building. Funding is used in the format of performance funding to promote more graduates in STEM. Inducements typically result in short-term gains that go away once the funding is eliminated, whereas funding for capacity building can result in longer term gains. In this case, changes to programming within Virginia's colleges can institutionalize changes that result in more graduates.

The processes underway with the 6-year plans in Virginia and with the Achieve 2015 process in VCCS in particular, may ultimately result in system changes (McDonnell & Elmore, 1987) or second-order change (Bartunek & Moch, 1987). The recent reengineering of developmental programming at the community colleges indicates a move toward deeper changes in the system. What remains unknown, however, is if these initiatives will have staying power and if changes will occur in other facets of the organization. Contributing to the early success of the implementation of TJ21 within the community college sector in Virginia was the change already underway.

Change requires the creation of a shared vision (Fullan, 2006; Kotter & Cohen, 2002). The processes in place to create Achieve 2015 built on planning efforts began in 2003. Establishing a track record of planning and creating processes to increase involvement of stakeholders meant that VCCS was in a different starting point to be able to implement the changes outlined in TJ21. An argument can be made that TJ21 merely provided a particular frame or saga to planning already underway and that the Act did not require the community colleges to fundamentally change their plans or actions. The announcement of President Obama's American Graduation Initiative and the introduction of Governor McDonnell's TJ21 legislation received little criticism as the message was communicated that there was too much at stake to ignore the need to invest in America's higher education system (The White House, 2009; Virginia Higher Education Opportunity Act, 2011). As outlined by Fowler (2009) above, the questions asked in creating policy about the rationale and appropriateness to adopt the new policy and support of stakeholders provide a means of evaluation. With respect to the TJ21 and its implementation in Virginia's community colleges, the responses to these questions show that there was a need to implement the policy given the shortages in STEM and in college graduates in the state and that stakeholders supported these goals.

The period of time following the onset of these initiatives is what Fowler (2009) referred to as the mobilization for implementation and it includes "policy adoption, planning, and the gathering of resources" (p. 285). The successful implementation of a new policy hinges on the mobilization phase because it establishes the case and foundation on which the implementation of the policy will be built, establishes the reason for the policy, and often can incentivize stakeholders to see it successfully adopted and implemented. According to Fullan (2006) and Fowler (2009), motivation is critical for effective educational reform and policy implementation. Davies' (2006) approach further supports and reinforces the need for buy-in from stakeholders when he describes the importance of "ground[ing an] agenda and its priorities in the needs of state residents" (p. iv), again making the connection between the policy and its intended outcomes. Creating a shared vision, ensuring that a strategic plan drives efforts, forming partnerships with a key stakeholders and leaders, and establishing a structure where partners are held accountable leads to successful implementation (Johnson, 2012). The leadership in VCCS supports the change efforts underway.

The VCCS has been operating under the strategic goals of Achieve 2015 for two years and continues to refine the strategies for reaching these goals on an annual basis. With the introduction of the TJ21 legislation just one year ago, the emphasis placed on STEM degree production created the opportunity for the VCCS to tweak the language in their established plan to include an intentional pursuit of new STEM-related academic programs. Slight changes in language allowed for a framing of the policy within an already existing structure. In this reference, the VCCS did not need to change their direction as they were able to easily incorporate the pursuit of STEM programs under the umbrella of Achieve 2015. With implementation strategies and evaluation measures already in the works, the VCCS was able to continue its course with Achieve 2015 and the redesign of developmental mathematics, making strides toward increasing the number of STEM graduates per the TJ21 legislation with only minor programmatic modifications.

However, there is more that can and should be done to ensure the attainment of the TJ21 STEM-related goals and performance incentives. As noted, the changes to date have primarily involved first-order change, versus deep second-order changes (Bartunek & Moch, 1987). According to the National Governor's Association (2011), policy gaps still need to be addressed so that improvements in building a STEM-skilled workforce can be realized. It is important that community colleges identify regional STEM-skill needs from businesses and gain a better understanding of the labor market demands; support new models of STEM skill development, such as career pathways, STEM early college high schools, earn and learn programs, and STEM bridge programs; and ensure STEM credits and degrees are transferrable (NGA, 2011). Given that only 48% of the community colleges in Virginia had specific reference to STEM in their planning documents may indicate that local needs are still being determined or that some regions have different requirements for STEM.

A sense of urgency for change was created with the passage of TJ21 (Kotter & Cohen, 2002), in particular for STEM. What remains unknown is how well the current change initiatives will become institutionalized within the community college system. The presence of a continuous planning cycle bodes well for attention to current challenges as a venue is present to accommodate external changes. While it is likely that the VCCS will continue to refine its practices to better serve the needs of the state and regional communities, future monitoring of performance will be necessary to ensure adequate progress is achieved and maintain appropriate accountability.

Other states can learn some central lessons from the experiences to date in Virginia. First, it is important to create a shared vision to support policies regarding STEM in community colleges. Incorporating a process to support STEM programming within an existing planning process is preferred to merely using STEM policy as an add-on. Second, using data to determine effectiveness of programming and to identifying community need is critical. Communicating with key stakeholders in the community, involving K-12 educators and employers, and framing the initiatives using similar language helps to institutionalize changes in policy. How community college leaders talk about changes to promote STEM matters. The ways in which leaders frame change on campus influences how campus members and community stakeholders understand what is going on and form their interpretations of activities and outcomes (Eddy, 2003). Finally, moving beyond first-order change requires a level of commitment to asking hard questions about past practices, how things are done, and what can be changed. This level of change requires top-level leader buy-in, constant communication, and recognition that nothing is sacred. True, deep change is hard. The analysis conducted on efforts in the state of Virginia indicates that the implementation of the new higher education policy may ultimately result in true change. Time will tell.

References

2007 Two-Year College Transfer Grant Program. (§ 23-38.10:9).
2011 Higher Education Opportunity Act of Virginia. (2011) (§ 23-38.87:10). (§ 23-38.87:19).
Bartunek, J, M., & Moch, K (1987). First order, second order, and third-order change and organizational development interventions: A cognitive perspective. *Applied Behavioral Science, 23*, 483-504.
Bowen, H. (1977). *Investment in learning: The individual and social value of American higher education.* San Francisco, CA: Jossey-Bass.
Breiner, J. M., Harkness, S. S., Johnson, C. C., & Koehler, C. M. (2012). What is STEM? A discussion about conceptions of STEM in education and partnerships. *School of Science and Mathematics, 112*(1), 3–11.
Bureau of Labor Statistics, U.S. Department of Labor. (2012). *Occupational outlook handbook, 2012–2013 edition: Projections overview.* Retrieved from http://www.bls.gov/ooh/About/Projections-Overview.htm#population
Business Higher Education Forum. (2010). *Modeling the role of community colleges in increasing educational attainment and workforce preparedness. BHEF working paper.* Washington, DC: Author.
Carnevale, A. P., Smith, N., & Melton, M. (2011). *STEM: Science technology engineering mathematics state-level analysis.* Washington, DC: Georgetown University Center on Education and the Workforce.
Carnevale, A. P., Smith, N., Stone, J. R., Kotamraju, P., Steuernagel, B., & Green, K. A. (2011). *Career clusters: Forecasting demand for high school through college jobs, 2008–2018. State data.* Washington, DC: Georgetown University Center on Education and the Workforce.
Clark, B. R. (1972). The organizational saga in higher education. *Administrative Science Quarterly, XVII,* 178-184.
Cohen, A. M., & Brawer, F. B. (2008). *The American community college* (5th ed.). San Francisco, CA: Jossey-Bass.
Davies, G. K. (2006). *Setting a public agenda for higher education in the states: Lessons learned from the national collaborative for higher education policy.* Denver, CO: The Education Commission of the States.
Douglass, J. A. (2007). The entrepreneurial state and research universities in the United States: Policy and new state-based initiatives. *Higher Education Management and Policy, 19*(1), 84–120.
DuBois, G. (2011, October). *Virginia's community colleges: Reengineering update.* Paper presented at the VCCS Administrative Services Fall Conference, Richmond, VA.
Eddy, P. L. (2003). Sensemaking on campus: How community college presidents frame change. *Community College Journal of Research and Practice, 27*(6), 453–471.
Fairhurst, G. T., & Sarr, R. A. (1996). *The art of framing: Managing the language of leadership.* San Francisco, CA: Jossey-Bass.
Fowler, F. C. (2009). *Policy studies for educational leaders: An introduction.* Boston, MA: Pearson Education.
Fullan, M. (2002). *Change forces with a vengeance.* New York, NY: Routledge.
Fullan, M. (2006). Change theory: A force for school improvement (Seminar Series Paper No. 157). Jolimont, Australia: Centre for Strategic Education.
Gonzalez, J. (2011, July). Virginia community colleges dive headfirst into remedial-math redesign. *Chronicle of Higher Education.* Retrieved from http://chronicle.com/article/Va-Community-Colleges-Dive/128430/
Hagedorn, L. S., & Purnamasari, A, V. (2012). A realistic look at STEM and the role of community colleges. *Community College Review, 40*(2), 145–164.
Harris, S. G. (1994). Organizational culture and individual sensemaking: A schema-based perspective. *Organization Science, 5*(3), 309–321.
Heidel, J., Ali, H., Corbett, B., Liu, J., Morrison, B., O'Connor, M., ... Ryan, C. (2011). Increasing the number of homegrown STEM majors: What works and what doesn't. *Science Educator, 20*(1), 49–54.
Henderson, C., Beach, A., & Finkelstein, N. (2011). Facilitating change in undergraduate STEM

instructional practices: An analytic review of the literature. *Journal of Research in Science Teaching, 48*(8), 952–984.

Huberman, M., & Miles, M. B. (1984). *Qualitative data analysis: An expanded sourcebook* (2nd ed.). Thousand Oaks, CA: Sage.

Johnson, C. C. (2012). Implementation of STEM education policy: Challenges, progress, and lessons learned. *School of Science and Mathematics, 112*(1), 44–55.

Keller, G. (1983). *Academic strategy: The management revolution in American higher education*. Baltimore, MD: The Johns Hopkins University Press.

Kotter, J. P., & Cohen, D. S. (2002). *The heart of change: Real-life stories of how people change their organizations*. Cambridge, MA: Harvard Business School Press.

Landon, M. G. (2009). *Emerging workforce trends and issues impacting the Virginia Community College system* (Doctoral dissertation). ProQuest LLC.

Lenaburg, L., Aguirre, O., Goodchild, F., & Kuhn, J. W. (2012). Expanding pathways: A summer bridge program for community college STEM students. *Community College Journal of Research and Practice, 36*(3), 153–168.

Lewis, J. (n.d.). Retrieved from http://www2.nr.edu/mall/

Lumina Foundation. (2010). *A stronger nation through higher education: How and why the nation must reach a "big goal" for college attainment*. Indianapolis, IN: Author.

Malcom, L. E. (2010). Charting the pathways to STEM for Latina/o students: The role of community colleges. *New Directions for Institutional Research, 148*, 29–40.

McDonnell, B. (2009). *Agenda: Higher education*. Retrieved from http://www.bobmcdonnell.com/index.php/issues/higher_education#Bob%27s%20Plan%20for%20Higher%E2%80%A6

McDonnell, L. M., & Elmore, R. F. (1987). Getting the job done: Alternative policy instruments. *Educational Evaluation and Policy Analysis, 9*(2), 133–152.

Minor, J. C. (2012). America's community colleges. *Science, 335*(23), 1409.

National Governor's Association. (2011). *Using community colleges to build a STEM-skilled workforce*. Washington, DC: Author.

Neumann, A. (1995). On the making of hard times and good times. *Journal of Higher Education, 66*(1), 3–31.

Northern Virginia Community College (NOVA). (2012). *SySTEMic solutions: Inspiring minds for advanced technology*. Retrieved from http://www.nvcc.edu/about-nova/partnerships/systemic/model.html

Obama, B. (2009, February). Remarks of President Barack Obama presented at address to joint session of congress, Washington, DC. Retrieved from http://www.whitehouse.gov/the_press_office/Remarks-of-President-Barack-Obama-Address-to-Joint-Session-of-Congress

Preparing for the Top Jobs of the 21st Century: The Virginia Higher Education Opportunity Act of 2011, Chapter 869, Virginia Acts of Assembly, Legislative Session (2011).

Reyes, M. E. (2011). Unique challenges for women of color in STEM transferring from community college to universities. *Harvard Educational Review, 81*(2), 241–263.

Rowley, D. J., Lujan, H. D., & Dolence, M. G. (1997). *Strategic change in colleges and universities: Planning to survive and prosper*. San Francisco, CA: Jossey-Bass.

State Council for Higher Education in Virginia (SCHEV). (2012). *Programs created since 2000*. Retrieved from http://research.schev.edu/completions/new_stem_progs.asp

State Higher Education Executive Officers. (2011). *Degree completion by program area: State by state reports (Virginia)*. Boulder, CO: Author.

Thayer, L. (1988). Leadership/communication: A critical review and a modest proposal. In G. M. Goldhaber & G. A. Barnett (Eds.), *Handbook of organizational communication* (pp. 231–263). Norwood, NJ: Ablex.

Townsend, B. K. (Ed.). (1999). Understanding the impact of reverse transfer students on community colleges. *New Directions for Community Colleges, 106*, 47–56.

Virginia Community College System. (2011). *Chancellor's 2011–2012 Goals Aligned Achieve 2015*. Retrieved from http://www.vccs.edu/WhoWeAre/Achieve2015.aspx

Virginia's Community Colleges (VCCS). (2012). http://www.vccs.edu/WhoWeAre/FastFacts.aspx

Virginia's Community Colleges (VCCS). (n.d.) Who we are: Our history. Retrieved from http://www.vccs.edu/WhoWeAre/OurHistory.aspx

Weick, K. E. (1995). *Sensemaking in organizations*. Thousand Oaks, CA: Sage.

The White House. (2009). *Building American skills by strengthening community colleges*. Retrieved from http://www.whitehouse.gov/issues/education/higher-education/building-american-skills-through-community-colleges

3

THE NEED FOR INTEGRATED WORKFORCE DEVELOPMENT SYSTEMS TO BROADEN THE PARTICIPATION OF UNDERREPRESENTED STUDENTS IN STEM-RELATED FIELDS

Victor Hernandez-Gantes and Edward C. Fletcher Jr.

Background

According to a recent Global Competitiveness Report, the United States has continued to lose its economic competitiveness and now ranks fifth in the world behind Switzerland, Singapore, Sweden, and Finland (World Economic Forum, 2011). In the midst of this increased global competition fueled by dramatic changes in technology, a popular storyline has emerged in the literature pointing to the need for a well-prepared science, technology, engineering, and mathematics (STEM) workforce if the United States is to preserve its economic competitiveness (Jobs for the Future, 2007; National Academy of Sciences, 2005, 2007). Although it has been projected that STEM occupations will represent only 5% of all jobs in the U.S. economy by 2018, STEM employment serves as a gauge of economic competitiveness, as it is directly tied to innovation and economic development (Carnevale, Smith, & Melton, 2011). To be sure, there is a growing concern for a balanced demand and supply of talent in STEM given the consistent reports from employers noting the absence of qualified workers in related fields (Carnevale, Smith, & Strohl, 2010). Further, there is also a growing concern for the decline in participation of minorities and women in STEM occupations, prompting a movement for broadening their participation in the related education pipeline (George, Neale, Van Horne, & Malcom, 2001; Kim, 2011).

A browsing of research and programmatic initiatives sponsored by the National Foundation (NSF) illustrates the interface of the above concerns as a means to promote an understanding of underlying issues and the nature and impact of potential solutions (Alvarez, Douglas, & Harris, 2010). These and other federal and state initiatives have pushed the STEM movement to the center stage of improvement efforts based on the premise of shortages of workers in related occupations and the lingering underrepresentation of minorities and women. While we find the underlying concerns to be valid, we note some problematic assumptions and offer an alternative analysis to the calls for improvement of the STEM pipeline and the broadening of student participation. In our view, at issue is that federal and state initiatives often assume the existence of a coherent system of workforce education and development where all the pipeline components are integrated. The reality is that we have a disjointed system involving four major components including K-12 education, community colleges, universities, and informal education in the workplace or community. The workforce system at the national level is designed to provide opportunities for training and retraining of the workforce providing states with the authority for customized implementation based on particular needs (Jobs for the Future, 2007). To this end, the many variations found in the system create a chaotic organization susceptible to political changes and blanket policies primarily viewing workforce development as a training tool largely disconnected to formal education (Grubb et al., 1999). In turn, the formal education pipeline is highly segmented preventing articulation agreements required for a more coherent system. Even in cases where segments of the educational system are articulated such as high schools and community colleges, the alignment with the labor market is often lacking (Ganzglass et al., 2000).

In this context, it is not surprising that efforts to broaden the participation of minorities and women in the STEM pipeline have yet to have a sustained impact. Clearly, there is a need for integrated workforce development systems engaging education institutions at all levels. Further, we need to go beyond broad STEM data to identify where the gaps are in specific disciplines, for particular groups, and based on local/regional needs. Building on these premises, the purpose of this chapter is to examine related literature supporting this call for the integration of workforce education systems as a means to broaden participation in STEM related fields. We begin with an overview of participation and degree completion trends, with particular focus on underrepresented groups and specific STEM fields. Next, we examine current initiatives seeking to broaden the participation of underrepresented groups in STEM career paths, followed by a discussion of workforce education and development issues. Finally, we explore the implications for an integrated workforce education and development system to engage communities and education entities in coherent efforts to broaden the participation of underrepresented groups in STEM education and work.

Participation Trends in the STEM Pipeline

The STEM designation has been used loosely to suggest a monolithic field when, in fact, it is a very diverse cluster representing a wide range of disciplines requiring a unifying operational definition. To this end, given the diversity and overlapping of occupations in each of the fields included in the acronym, there is no standard definition for the STEM cluster other than inclusion of fields such as science, technology, engineering, and mathematics. For example, the National Science Foundation (NSF) also includes social/behavioral sciences such as psychology, economics, sociology, and political science as part of its STEM definition (Green, 2007). Albeit this loose understanding, federal initiatives often refer to STEM as education in the natural sciences (e.g., physical and biological sciences), technologies (e.g., computer/information sciences), engineering, and mathematics (Kuenzi, Matthews, & Mangan, 2006; National Governors Association, 2007). Further, reports on STEM-related topics produced by the National Center for Education Statistics (NCES) typically use the latter designation embraced by federal initiatives (NCES, 2009, 2011).

With an understanding of this caveat in definition, it is important to examine participation trends in general, the storyline about underrepresentation of ethnic minorities and women, and how related issues play out in the context of particular STEM disciplines, and geographical regions.

General Participation Trends

Participation trends in the STEM pipeline have been well documented in recent years providing the fodder for reported shortages of scientists and engineers. For example, based on a national sample of undergraduate students, the National Center for Education Statistics (2009) reported that, of all students enrolled in U.S. postsecondary institutions in 2003–2004, only about 14% were enrolled in a STEM fields. Further, it has been noted that although the enrollment in STEM degree fields increased between 1994–1995 and 2003–2004, the percentage of all undergraduate degrees awarded in STEM fields declined from 32% to 27% (Consortium of Social Sciences Associations [CSSA], 2008). This level of talent supply in STEM fields is often viewed as inadequate in meeting the demand for related workers and professionals in the labor market (Commission on Professionals in Science and Technology [CPST], 2007; Lowell & Regets, 2006).

However, more recent data on STEM awards including degrees and certificates helps clarify the storyline about the imbalance on the supply of related talent (NCES, 2011). While data from 2000–2001 and 2008–2009 showed even lower rates of STEM awards at 12.9% and 10.7%, respectively, a breakdown by STEM discipline suggests uneven trends. That is, most positive gains in STEM awards have been observed in science technologies, biological sciences, and mathematics and to a much lesser extent in engineering. In computer science though, the percent change for the same period was a negative 14.7%. Moreover, when breaking down the STEM degrees awarded by level, it is revealed that for the same period between 2000–2009, most gains were observed at the advanced doctoral level (45.9%) with diminishing gains at the master (32.3%) and bachelor (19.9%) levels, while reaching negative changes at the associate's degree level (-1.7%). Further examination of percent change in certificates awarded during the same period suggests that technician preparation is where the problem lies with negative changes in the awarding of certificates involving less than 1 year (-23.9%), 1 year but less than 2 years (-36.3%), and 2 but less than 4 years (-.9.9%) of study (NCES, 2011).

Participation of Minorities and Women

In terms of participation of minorities and women as measured by number of STEM degree awards, by all accounts, these groups are lagging behind significantly when compared to the White majority (Carnevale et al., 2011). For example, during the 2000–2009 period, the percentage of awards earned by Whites hovered around 60%, while minority groups remained relatively stable at rates ranging from 6% (American Indian) to 9.5% (Asian). Likewise, the STEM field continues to be dominated by males, with women receiving only about 32% of all awards during the same period (NCES, 2011). Based on data like this, a movement for broadening the participation of underrepresented groups has gained momentum in recent years as demonstrated by federal initiatives in STEM fields (de los Santos, Keller, Nettles, Payan, & Magallan, 2006; George et al., 2001; NSF, 2012).

Although the overall data supports the call for increased participation of minorities and women, it is also helpful to examine emerging trends by STEM discipline to pinpoint where the problematic areas are located. According to NCES (2011), from 2000 to 2009, positive changes have been observed for all minority groups and women in engineering, biological sciences, mathematics, physics, and science technologies. The only exception is computer science where the percentage of awards has decreased across the board with the most impact on Asians (-45.5%), African Americans (-10.5), and women (-43.7%). Similarly, mirroring the overall trends when breaking down awards data by degree level, positive gains are observed between 2000 and 2009 at the bachelor and graduate level, while lower to negative change is observed at the associate's degree level for minority groups with women being the most affected (-23.9% change). This trend is further confirmed when examining the percent change in STEM-related certificates earned during the same period by minorities and women with Asians being most affected at all levels (from less than 1 year to between 2 and 4 years of study), followed by African Americans, and Hispanics. At the certificate level, women experienced the worst decline ranging from -12.8% change after 2 but less than 4 years of study, to -53.8% with those individuals completing at least 1 but less than 2 years of study (NCES, 2011).

An Alternative Analysis to the Popular Storyline

The popular storyline found in the STEM literature and federal initiatives has suggested a shortage of talent in the educational pipeline. However, the examination of STEM-related data on participation and degrees awarded leads to two important points. First, it is important to break down data by specific discipline for meaningful interpretation of trends in the field to understand where in the education pipeline problems are more acute and for a more useful translation to policymaking. As noted above, engineering in general and computer science in particular are the two broad disciplines where more attention is needed. Also, it appears technician preparation is the level where attention is most needed when comparing undergraduate and graduate levels. Second, underrepresentation of minorities and women remains a problem in general, although it appears to be even more problematic in computer science and in technician preparation where representation continues to erode. Further, these trends may not fully account for the growing Latino population whose representation in the pipeline might be underestimated and misunderstood (Chapa & De La Rosa, 2006; de los Santos et al., 2006).

Other researchers have also challenged the popular storyline about shortages in STEM fields and warnings about the erosion of the nation's economic competitiveness (Black & Stephan, 2005; Hall, Dickerson, Batts, Kauffmann, & Bosse, 2011; Metcalf, 2011). They argue that an alternative view often overlooked in this debate is that the imbalance of STEM workers should be evaluated locally rather than nationally. That is, it is projected that STEM occupations will grow much faster (17%) than the general economy (10%) by 2018 (Carnevale et al., 2011). However, most STEM jobs are often found on the East and West coasts and are not as available in the Midwest and South. Mather (2006), for example, reported that given the differential STEM employment opportunities, as states compete for high-tech workers, the geographical distribution of talent changes, following the flow of skill demands.

The geographical differential in STEM-related job opportunities is important to note because it points to the interface of economic development, education, and policymaking. As states seek to boost the growth of STEM related economies, it is critical to take into consideration gaps in specific STEM disciplines and the alignment with educational preparation. Further, the geographical concentration of minorities may interact with the distribution of STEM employment opportunities boosting or hindering their representation depending on their location (Carnevale et al., 2011; Mather, 2006).

Thus, based on participation trends by STEM field, we offer an alternative view on the popular notion that there is a shortage of qualified workers in STEM-related areas and there are not enough graduates to fulfill available jobs. Our view is that it depends. It depends on particular STEM disciplinary cluster, educational level, and geographical region. By all accounts, computer science appears to be the STEM discipline most affected by a declining participation trend. In turn, the participation of women in engineering and computer science has been most affected as well. Further, it is also critical to pinpoint where the STEM sectors are located geographically to target initiatives in specific states taking into consideration the geographical concentration and growth rate of minority groups.

Fixing the STEM Pipeline: What Are We Doing About It?

Given the popular storyline on the status of STEM participation outlined above, and related concerns about the implications for global competitiveness of the United States, a national call for action emerged in the 2000s. In 2005, leading scientific organizations in the country including the National Academy of Science, the National Academy of Engineering, and the Institute of Medicine, released a report—which was further revised in 2007—to spearhead initiatives in response to shortages of STEM scientists and workers and to strengthen the entire STEM educational pipeline (National Academy of Sciences, 2007). Specifically, the report recommended boosting K-12 STEM education by enhancing the teaching capacity, increasing support for STEM research and innovation as the means for creating STEM jobs and related economic development, and broadening student participation in the field. These recommendations catalyzed the current emphasis of initiatives for improvement of the STEM educational pipeline.

Enhancing Teaching Capacity

The first recommendation advocated by the report was to improve STEM education by increasing the K-12 teaching capacity in America's schools through three major actions: (a) annual recruitment of 10,000 science and mathematics teachers, (b) a program of professional development for 250,000 STEM teachers, and (c) increasing the number of students passing Advanced Placement (AP) and International Baccalaureate (IB) science and mathematics courses. The underlying premise of these strategies is the potential multiplication impact of each teacher on about 1,000 students over the span of a teaching career (National Academy of Sciences, 2005, 2007). However, at a time when the teaching profession is being highly scrutinized, the goal of recruiting and preparing 10,000 science and mathematics teachers annually appeared to be quite ambitious. In addition, aside from being overly ambitious, at issue was who would set these actions in motion.

Given the more problematic nature of student participation in computing education, the Computing Education community responded to this call with a more measured approach. As noted above, student participation in computing has been on the decline over the past decade, especially among minorities and women (NCES, 2011) even though job opportunities in this field are on the rise (American College Testing [ACT], 2010). To address the problem, the Computing Education community launched an initiative supported by the National Science Foundation, the Computing Education for the 21st Century

(CE21), CS 10K, with the goal of recruiting 10,000 high school teachers and engaging them in professional development to teach new computing curriculum in 10,000 schools by 2015 (Cuny, 2010). The strategy of this community is to concurrently develop a middle school curriculum and a new AP high school course, which will impact early undergraduate education as well. Under this approach, the teacher recruitment and preparation strategy along with the curricular development approach appears promising. The CE21 initiative also emphasizes school practices and supports with potential application to broadening the participation of students in computing and for successful transitions to work or further education in computing-related fields (CE21, 2012).

Promoting STEM Research and Innovation

Two additional recommendations from the joint report of national organizations were characterized as "sowing the seeds" for innovation through long-term basic research as the basis for maintaining global competitiveness (National Academies of Sciences, 2007). The report proposed specific actions including an increase in federal investment to boost interdisciplinary STEM research and the work of junior researchers as a companion action to enhance research capacity. Further, the report advocated for investments in research infrastructure to support innovation in STEM areas, and for promoting creative research in areas of national interest such as energy.

The proposed actions were certainly timely and warranted considering that the U.S. global competitiveness has experienced a slow but consistent decline in recent years, now ranking fifth in the world (World Economic Forum, 2011). To this end, the role of technology and innovation is viewed as essential for economic competitiveness in a global economy. Thus, investments in related research, human capital, and supporting infrastructure are critical for enhancing and sustaining the research and development capacity of a country (Nelson, 2002; OECD, 2010; World Economic Forum, 2011). In the United States, the National Science Foundation has served as the coordinating vehicle for the promotion of innovation through the support of basic research and prototype development (NSF, 2010). In this regard, NSF has followed through with programs that take into consideration the innovation ecosystem involving researchers and research infrastructure (Foray 2009; NSF, 2010). Examples of such programs include the support of Engineering Research Centers (ERC) and Partnerships for Innovation (PFI) as a means to promote basic research through collaborative partnerships. Further, NSF promotes the development of innovation through industry and innovation fellowships, early researcher awards, researcher networks, and many other activities and programs (NSF, 2010).

Ultimately, as the bulk of innovation work is carried out at research universities, it is expected that a stronger innovation ecosystem could contribute to the STEM education pipeline through graduates with the capacity to sustain the U.S. global competitiveness. In this context, a companion recommendation of the joint report published by the National Academy of Sciences (2007) called for incentives to diffuse innovation through manufacturing and marketing, a focus on high-skill/high-paying jobs, and economic policies to ensure alignment of demand-supply of talent as well (Nelson, 2002).

Broadening STEM Participation

The call to broaden student participation in higher education has been an initiative that has received more attention in recent years. Specifically, the recommendation made in the report published by the National Academy of Sciences (2007) suggested developing, recruiting, and retaining top students in the country to increase the number of graduates with undergraduate STEM degrees. In addition, the joint report also called for increasing the number of graduate students in areas of particular national interest, and attracting top international talent to broaden participation in STEM fields.

Again, taking heed of the declining participation in computing, the Computing Education community is an example of a concerted response to broadening participation in the field. To illustrate, the NSF's CE21 Program sponsors research and demonstration projects focusing on teaching and learning, and on engagement and retention of students who are underrepresented in the computing education pipeline from K-12 to graduate level. This program provides for opportunities to research, demonstrate, and scale up promising interventions. In general, the CE21 program seeks to broaden the participation of diverse student groups in computing while increasing their job opportunities in the field (CE21, 2012; NSF, 2012).

Making Sense of Current Initiatives

In general, it is undisputable that global competitiveness is closely tied to STEM research and innovation, which—with appropriate incentives—can contribute to continued economic development. In turn, the only way to translate new knowledge and skills into the STEM pipeline is through enhanced teaching capacity at all levels, and with particular emphasis at the secondary level. Ultimately, the goal is to attract and engage a broader student population in STEM education if we are to increase the diversity in the STEM workforce. To this end, albeit the promising nature of current initiatives, we point out some caveats for successful implementation.

To begin, research and innovation is typically associated with research universities engaged in partnerships with industries that cluster and concentrate in particular geographical areas. Thus, pockets of innovative communities have been created, for example, in the Silicon Valley in California and the Research Triangle in North Carolina. Thus, one overriding concern is the potential for preserving the comparative advantage in research and innovation of communities that are already well established. This, in turn, may only reinforce the issues related to demand and supply of the talent discussed above. Further, although a boost in research and innovation may have some direct impact for teaching and learning in higher education as faculty fulfills research and teaching roles, enhancing teaching capacity at the secondary level may be problematic. Teacher preparation is largely disconnected from research, focusing primarily on general disciplinary content (e.g., mathematics, science) and teaching methods, while in the case of engineering and technology there are virtually no specific programs. As a result, the teacher workforce for STEM disciplines—other than mathematics and science, may exhibit a wide range of qualifications for teaching engineering and computer science courses, for instance (Association for Computing Machinery [ACM], 2010; Computer Science Teacher Association, 2008). In this context, teacher recruitment and professional development represent important challenges for fulfilling related efforts to enhance teaching capacity.

Regarding efforts for broadening the participation of students in STEM fields, as noted earlier, it is important to go beyond the blanket call and into an understanding of participation trends by particular groups and by field. To this end, aside from teacher certification issues, the NSF's CE21 program illustrates a concerted approach by clearly responding to the calls for enhancing the STEM pipeline through investments in teaching capacity, curriculum development, and targeted student participation (ACM, 2010; Cuny, 2010; NSF, 2012). What is generally missing in the calls for enhancing the STEM pipeline, is the role of career and technical education in technician preparation at the high school and community college level. Although the calls for improvement are often prefaced by the need for a well-educated workforce in a global economy, the connection to career and technical education is often overlooked (ACM, 2010; National Academy of Sciences, 2007).

For example, despite the fact that career academies have been found to have an impact on student engagement and program completion—especially for minority groups (Castellano et al., 2007), these programs are often ignored in STEM discussions. Moreover, it has been documented that the majority of growth in STEM jobs require technician preparation involving between 1 and less than 4 years of study leading to a certificate or an associate's degree (Carnevale et al., 2011). In addition, enrollment trends indicate that participation in STEM-related certificate and associate's degree programs are where the largest declines are observed (NCES, 2011). Yet, career and technical education at the secondary level and technician preparation at the community college level are often viewed as an afterthought despite their relevance and the opportunities for articulation in the STEM pipeline between academic and technical education and institutional levels.

To summarize, we argue that it is imperative to view STEM research and innovation as a strategy for economic development with direct implications for workforce education within a geographical context. Further, investments in teaching capacity have to acknowledge the particular caveats of teacher preparation and qualification requirements of STEM disciplines. In turn, initiatives to broaden the participation of STEM students, should take into consideration the geographical context, gaps in specific STEM fields, and trends in particular educational levels. On the latter, the integration with career and technical preparation may represent an opportunity for enhancing the pool of students in the STEM pipeline at the secondary and postsecondary level (e.g., dual enrollment).

The Need for an Integrated Workforce Development System

Studies on the STEM workforce have often relied on a demand/supply model to determine and/or predict shortages in the context of the education pipeline. This model was introduced by the National Science Foundation in the 1970s and served as the basis for policy decisions and education initiatives in the 1980s as a response to perceived international technological competition (Lucena, 2005; Metcalf, 2007). This model, still in use today, assumes a workforce education and development system (or pipeline) functioning as a series of interconnected components contributing to the preparation of scientists and engineers. Concurrently, this model has been used to quantify the output of completers moving from secondary to higher education, and to work in STEM related areas (National Research Council, 1986; Metcalf, 2010).

Despite many issues with the definition of STEM as an occupational cluster and head counting, the pipeline model has remained as the primary vehicle for quantifying participation in STEM fields (Lucena, 2005; Metcalf, 2007; Teitelbaum, 2003). As such, an element that has been largely ignored in the analysis of the pipeline model is the assumption of integrated linearity. That is, the view of a STEM education pipeline takes for granted the existence of a workforce education and development system under the assumption that all its formal and informal education components are coherently integrated at the national, state, and local level. An analysis of this established view is particular important to understand the implications for initiatives designed to boost and broaden the participation in STEM fields.

Shift from Spare Parts to a System Approach

As commonly understood, a system is made up of discrete parts designed to work together to produce a shared outcome. In this context, a major assumption of a system is that all components are holistically integrated to produce a shared result. Without integration, there is really no system but a collection of spared parts producing parallel rather than shared results. As such, systems theory offers a framework for understanding the complementary connections between individual components and the mediating factors that make them work as a whole (Adelman & Taylor, 2003; Chen & Stroup, 1993). In simplistic terms, the premise of a general systems theory is that the whole is more than the sum of the parts, requiring component integration to produce a shared goal.

To this end, the STEM pipeline model and related initiatives for improvement represent a reaction to the changing nature of work and the new skills requirements in the labor market. The expected results though, assume an education system that is fully integrated and responsive to the emerging skills requirements and the needs of the STEM labor market. According to the U.S. Department of Labor (2012), the workforce system is a network of federal, state, and local offices working together with employers, educators, and community leaders to promote economic development in regional economies through talent development. In reality, although the public workforce system recognizes that education and training for individuals must align with the needs of business and industry, most of the emphasis is actually on workers' training, rather than a comprehensive system of education and training.

In the United States it is argued that we don't really have a coherent workforce system accounting for economy-work-education connections. When we refer to the system of "workforce education" or "workforce development" we may get a variety of perspectives on what the terms mean. To be sure we have system components supported by national legislation with funding tied to specific programs based on shared formulas with the states, which in turn can push their own ideas to ensure workforce competitiveness. Thus, what we have is a collection of programs involving a mix of formal and informal provisions to prepare youth and adults for and through work including technical education, job training, adult education, targeted education and training for specific groups, and on-the-job training provided by employers. Thus, there are multiple entry points, changing purposes and overlaps shaped by the politics of the day describing a chaotic organization rather than a coherent system (Grubb et al., 1999).

In 1999, Grubb reported an assessment that still stands today. He described a three-track system comprised of formal education, workforce development, and technical education. In the United States, the education component includes schools, colleges, and universities with each institutional level sharing recognizable characteristics and embracing academic and relatively broad conceptions of career preparation. In turn, the workforce development system in place today, as noted above, emphasizes occupational preparation through a variety of job-specific training programs or on-the-job training. In this regard, workforce development is somewhat connected to postsecondary technical education but such connections vary greatly depending on shifting dynamics shaped by state and local views. The third track is formal technical education provided in high schools, community colleges, and technical institutes—which in some cases can serve as a bridge between academic education and workforce development (Patton, 2008). In recent years, the "bridging" role of technical education has received more attention due to positive results of high school career academies featuring the integration of academic and technical education with certification requirements (Castellano et al., 2007; Kemple, 2008). In turn, the role of community colleges in workforce development has received greater attention as for their potential contributions to both education and training (Austin, 2012; Patton, 2008; Phelps, 2012). Hence, although there are some potential bridges in the system, the academic track functions largely in isolation, while the connections between workforce development and technical education varies greatly depending upon local circumstances. Given this organizational picture of workforce education and development, the question is: Is it possible to implement successful STEM initiatives under such conditions?

Toward an Integrated System: What Is Possible?

Although well intentioned, the problem with initiatives to improve the STEM education pipeline is that proposed improvements are often tied to formal academic education, and to a much lesser extent to technical education in secondary and postsecondary settings. In terms of connections to established workforce development structures and secondary technical education, such alignment is virtually absent in related improvement discussions. To this end, the proposed STEM initiatives do not represent system solutions as they primarily focus on the academic track although the larger gaps in the STEM pipeline are in technician education involving secondary and postsecondary technical education (Carnevale et al., 2010; NCES, 2011). Further, even in the academic education track, vertical integration between education levels is assumed, when in fact high schools, community colleges, and universities work in relative isolation. In addition, the premises of STEM improvement initiatives are often disconnected of the regional labor markets as they may be based on broad STEM data rather than an analysis of disciplinary clusters, and educational levels in the context of specific geographical regions. Thus, if we are to adjust the STEM pipeline to the realities of the emerging skills demands, there is a need for a workforce education and development system involving vertical integration within academic education as well as holistic integration with the established workforce development and technical education components.

Large organizational systems that have a longstanding record of development are slow to change though. The reality is that we have a complex education system shaped by social, economic, and political factors with each state following its own approach (Grubb, 1999). Under these conditions, an integrated workforce education and development system represents an idea that may take decades to debate and resolve. Thus, the focus should be on what is possible to change based on a systems approach at the state, regional, or local level. For example, since states follow their own approaches, state and regional systems represent an opportunity for integration of efforts. This is supported by the fact that over the past decade, state governments have joined the bandwagon to improve the STEM pipeline through higher graduation requirements in mathematics and science, high school pre-engineering curricula, teacher training and recruitment, and dual enrollment in STEM courses among other measures (Zinth, 2006).

Some Possibilities for System Integration

Based on our review of participation trends and current initiatives, our view is that opportunities for integration are needed and possible in the form of systems approaches to addressing supply and demand of talent around specific STEM-related industry clusters. That is, how can stakeholders interested in advancing STEM education develop meaningful and mutually beneficial partnerships based on a systems approach? For specific STEM fields, which components of academic and technical education can be aligned to address labor demands in a regional workforce education system? We suggest three broad strategies for systems integration: Conducting sector analyses to determine specific occupational needs, articulating academic and technical education, and integrating career education to broaden the participation of students in STEM fields. The backdrop of these suggestions is the interplay of specific STEM-related industries and academic/technical education programs in regional context. Ultimately, workforce development systems have to make sense locally and become articulated at that level.

Conducting a sector analysis is essential for identifying specific labor demands in specific industry contexts including the determination of shortages in particular occupations. This analysis also serves as the platform for understanding the nature of occupations with labor deficits and for identifying the academic and technical preparation required for productive participation (Klaffke, 2012). Further, once labor demands are determined, STEM preparation should be viewed from the standpoint of a regional system for workforce development where occupational demands and STEM education supply are addressed by education-and-industry partners to promote enhanced articulation (Jobs for the Future, 2007; Sommers, 2009; Weeks, 2009). Unfortunately, information about the alignment of occupational demands and relevant education is not readily available in most communities. There are practically no structures in place to generate specific occupational information broken down by industry clusters to map out the regional occupational demand and supply. Thus, developing related understandings in the context of a regional system of workforce education for a given industry cluster, represents a potentially important strategy for advancing STEM preparation.

In this context, a certain level of alignment should be expected between occupational demand for STEM skills and the supply of technicians prepared in the STEM pipeline of secondary and postsecondary education through articulated pathways. In this regard, it is critical to determine the extent to which workforce education in a region is working congruently in meeting occupational demands in a particular industry. Are academic and technical education programs working in isolation? What programs and articulation structures are providing occupational paths to specific industry clusters? Who participates and what is the nature of participation and transitions to further education or work? These are the core questions we should ask to determine whether education-and-work components are functioning as a regional system (Hernandez-Gantes & Reeves, 2010).

In addition, it is also important to understand how STEM education integrates career education among high school, community college, and undergraduate programs to promote and sustain student participation (Carnevale, Strohl, & Smith, 2009; Hernandez-Gantes & Reeves, 2010; Weeks, 2009). This is an area some researchers have referred to as the pipeline "leaks" alluding to problems ensuring transitions from secondary to postsecondary education, particularly for minorities and women (Blickenstaff, 2005; Chapa & De La Rosa, 2006; Metcalf, 2007; Morris & Lee, 2004). Part of the problem is that STEM education is often considered as an academic issue primarily focusing on mathematics and science in secondary and postsecondary education. As such, this strategy ignores career and technical education as a legitimate STEM pathway even though programs like career academies have produced promising results regarding academic and occupational preparation with particular impact on minorities and women (Kemple, 2008). To broaden the participation of minorities and women in STEM fields, it is also important to bridge career and technical education and emphasize career guidance in middle and high schools to promote career awareness as a means to attract and retain students in the STEM pipeline (Metcalf, 2007; Xie & Shauman, 2003).

Conclusions and Implications for Broadening Student Participation

The focus of this chapter was to examine the STEM shortage from a more holistic and systemic perspective. To that end, the need to develop a national workforce development and education system was discussed. More specifically, this chapter first reviewed the literature with regard to participation and degree completion trends. In that respect, it was found that the commonly used STEM term is not entirely reflective of the wide array of disciplines within the STEM umbrella. And, this has resulted in common misconceptions regarding national participation trends. However, when analyzing each discipline within STEM, it was revealed that large disparities exist within engineering and computer science sectors, particularly among women at the sub baccalaureate levels. Second, this chapter discussed disconcerted efforts at the K-12 levels to increase the awareness and preparation of future STEM job incumbents through school reform initiatives such as career academies and Project Lead the Way. Third, this chapter contended that better articulation is needed to develop a pipeline of talent through the development of a stronger workforce development and education system—one that coherently connects K-12 schooling, community colleges, and universities with established partnerships with businesses and industry to address local, regional, state, and national economic and innovative development challenges.

Implications for Broadening Student Participation

There is no question STEM occupations are key to national economic competitiveness due to the connections to innovation and economic growth. To this end, as states and communities seek to boost the STEM workforce, we posit the following implications for increasing and broadening the participation in the STEM pipeline. First, we start with the obvious implication stemming from our analysis: The nation must establish a system for understanding STEM occupational demands in a regional context in order to determine occupational needs as tied to STEM fields. As some studies have suggested, the STEM supply problem is more than the hype for more scientists, engineers, and mathematicians. What is actually needed is more qualified STEM technicians in industries that cluster in different regions. To this end, it has been reported that the majority of STEM jobs require some postsecondary education or training rather than a college degree (Carnevale et al., 2011; Carnevale et al., 2009). Second, a companion strategy should be the identification of academic and technical education available in a regional area aligned with occupational STEM demands in specific fields. What secondary and postsecondary programs are available? Who participates and what rates? Where are the program and participation gaps? These questions are critical for promoting coherent participation and broadening the diversity in the STEM pipeline (Hernandez-Gantes & Reeves, 2010).

A third implication involves program articulation of academic and technical programs at the secondary and postsecondary level in terms of curricular pathways as well as transitional requirements and supports to address gaps and leaky points in the STEM pipeline. This will require articulation agreements between high schools and community colleges as well as among community colleges and four-year universities. This strategy will also require establishing compacts and advisory committees at the K-12, community college, and university levels to engage and partner with business and industry to evaluate and improve curricula as well as to connect local and regional workforce needs to relevant educational programs (Austin, 2012; Phelps, 2012). In turn, another important implication is the integration of career information in the workforce education system. Why is this important for minorities and women? The lack of specific occupational and wage information preserves student diversion to the idea that there is only one way to win; that a college degree is the only ticket for success.

For minority groups, such as Hispanics who are less likely to attend college and lag behind other ethnic groups in receiving college degrees, this idea prevents them from considering other pathways for success (Gándara, 2006; Grey, 2010; Hernandez, 2006). For women, it may prevent consideration from participation in STEM fields traditionally dominated by males (Carnevale et al., 2011). In this regard, it is imperative to create awareness as early as middle school about other ways to win integrating academic and technical education. For instance, students should be aware of program options that integrate academic and technical education curricula, such as career academies and magnet schools featuring STEM themes; and of opportunities for dual enrollment allowing high school students to earn college credit. As it is becoming increasingly evident in recent years, it is relevant occupational skills—not degrees—that provide an advantage in the STEM workforce (Carnevale et al., 2011; Gray, 2010).

To be sure, efforts to broaden the participation of women and minorities in STEM fields have been ongoing for over three decades, and although some promising efforts have emerged, systemic improvement continues to be a national challenge (Alvarez et al., 2010; CE21, 2012). Thus, in our view, unless we sort out STEM occupations and labor trends by field and in regional context, and establish partnerships to integrate and coherently articulate regional workforce education systems bridging academic and technical education, increasing and broadening student participation will remain unresolved.

References

Adelman, H., & Taylor, L. (2003). On the sustainability of project innovations as systemic change. *Journal of Educational & Psychological Consultation, 14*, 1–25.

Alvarez, C. A., Douglas, E., & Harris, B. (2010). STEM specialty programs: A pathway for underrepresented students into STEM fields. *NCSSSMST Journal, 16*(1), 27–29.

American College Testing. (2010). *The condition of college and career readiness 2010*. Iowa City, IA: Author.

Association for Computing Machinery. (2010). *Running on empty: The failure to teach K–12 computer science in the digital age*. New York, NY: Author.

Austin, J. (2012). Reanimating the vital center: Challenges and opportunities in the regional talent development pipeline. *New Directions for Community Colleges, 157*, 17–28.

Black, G. C., & Stephan, P. (2005). Bioinformatics: Recent trends in programs, placements, and job opportunities. *Biochemistry and Molecular Biology Education, 33*(1), 58–62.

Blickenstaff, J. C. (2005). Women and science careers: Leaky pipeline or gender filter? *Gender and Education, 17*(4), 369–386.

Carnevale, A. P., Smith, N., & Melton, M. (2011). *STEM: Science, technology, engineering, mathematics*. Washington, DC: Georgetown University, Center on Education and the Workforce.

Carnevale, A. P., Smith, N., & Strohl, J. (2010). *Help wanted: Projections of jobs and education requirements through 2018*. Washington, DC: Georgetown University Center on Education and the Workforce.

Carnevale, A. P., Strohl, J., & Smith, N. (2009). Help wanted: Postsecondary education and training required. Occupational outlook for community college students, *New Directions for Community Colleges, 146*, 21–31.

Castellano, M., Stone, J., Stringfield, S., Farley-Ripple, E., Overman, L., & Hussain, R. (2007).

Career-based comprehensive school reform: Serving disadvantaged youth in minority communities. St. Paul, MN: National Research Center for Career and Technical Education.

Chapa, J., & De La Rosa, B. (2006). The problematic pipeline: Demographic trends and Latino participation in graduate science, technology, engineering and mathematics programs. *Journal of Hispanic Higher Education, 5,* 203–221.

Chen, D., & Stroup, W. (1993). General systems theory: Toward a conceptual framework for science and technology education for all. *Journal for Science Education and Technology, 2*(3), 447–459.

Commission on Professionals in Science and Technology. (2007, October 9). *Is US science and technology adrift?* STEM Workforce Data Project: Report No. 8. Washington, DC: Author.

Computer Science Teacher Association. (2008). *Ensuring exemplary teaching in an essential discipline: Addressing the crisis in computer science teacher certification.* New York, NY: Association for Computing Machinery.

Computing Education for the 21st Century. (2012). *Computing portal: Connecting computing educators.* Retrieved July 6. 2012, from http://www.computingportal.org/node/3886

Consortium of Social Science Associations. (2008). *Enhancing diversity in science: A leadership retreat on the role of professional associations and scientific societies: A summary report.* Washington, DC: Author.

Cuny, J. (2010). Finding 10,000 teachers: Transforming high school computer science. *CSTA Voice, 5*(6), 1–2.

de los Santos, A.G., Keller, G.D., Nettles, M.T., Payan, R., & Magallan, R.F. (2006). Latino Achievement in the sciences, technology, engineering, and mathematics. *Journal of Higher Education, 5*(3), 200–202.

Foray, D. (2009). *The new economics of technology policy.* Northampton, MA: Elgar Publishing.

Gándara. P. (2006). Strengthening the academic pipeline leading to careers in math, science, and technology for Latino students. *Journal of Hispanic Higher Education, 5,* 222-237.

Ganzglass, E., Ridley, N., Simon, M., King, C. T., Narver, B. J., & Van Horn, C. (2000). *Building a next generation workforce development system.* New York, NY: Ford Foundation.

George, Y.S., Neale, D. S., Van Horne, V., & Malcom, S. M. (2001). *In pursuit of a diverse science, technology, engineering, and mathematics workforce: Recommended research priorities to enhance participation by underrepresented minorities.* Washington, DC: American Association for the Advancement of Science.

Green, M. (2007). *Science and engineering degrees: 1966–2004 (NSF 07-307).* Arlington, VA: National Science Foundation.

Grey, K. (2010, July). *Secondary and postsecondary career and technical education: Solving the quiet workforce development in Georgia and the nation.* Presentation conducted at annual meeting of the Georgia Association for Career and Technical Education, Atlanta, GA.

Grubb, W. N. (1999). From isolation to integration: Occupational education and the emerging systems of workforce development. *NCRVE CenterPoint, No. 3,* March. Berkeley: National Center for Research in Vocational Education, University of California.

Grubb, W. N., Badway, N., Bell, D., Chi, B., King, C., Herr, J., ... Taylor, J.C. (1999). *Toward order from chaos: State efforts to reform workforce development systems.* Berkeley: National Center for Research in Vocational Education, University of California.

Hall, C., Dickerson, J., Batts, D., Kauffmann, P., & Bosse, M. (2011). Are we missing opportunities to encourage interest in STEM fields? *Journal of Technology Education, 145,* 229–245.

Hernandez, V. M. (2006). Discovering other ways to succeed. *The Hispanic Outlook in Higher Education, 16,* 84.

Hernandez-Gantes, V. M., & Reeves, K. (2010). *Preparing medical device industry technicians in a regional workforce development system.* Grant proposal submitted to the National Science Foundation, Advanced Technology Education Program.

Jobs for the Future. (2007). *The STEM workforce challenge: The role of the public workforce system in a national solution for a competitive science, technology, engineering, and mathematics workforce.* Washington, DC: U.S. Department of Labor, Employment and Training Administration.

Kemple, J. J. (2008). *Career academies: Long-term impacts on labor market outcomes, educational attainment and transitions to adulthood.* New York, NY: MDRC.

Kim, Y. M. (2011). *Minorities in higher education: Twenty-fourth status report, 2011 supplement.* Washington, DC: American Council on Education.

Klaffke, H. (2012, July). *Development of occupational standards: Work process analysis as a tool to shape occupational standards.* Presentation conducted at the University of Bremen, Germany.

Kuenzi, J., Matthews, C., & Mangan, B. (2006). *Science, technology, engineering, and mathematics (STEM) education issues and legislative options.* Congressional Research Report. Washington, DC: Congressional Research Service.

Lowell, B. L., & Regets, M. (2006, August). *A half-century snapshot of the STEM workforce, 1950–2000.* Washington, DC: Commission on Professionals in Science and Technology.

Lucena, J. (2005). *Defending the nation: U.S. policymaking to create scientists and engineers from Sputnik to the "war against terrorism."* Boulder, CO: University Press of America.

Mather, M. (2006). *Is there a U.S. shortage of scientists and engineers?* Washington, DC: Population Reference Bureau.

Metcalf, H. (2007). *Recruitment, retention, and diversity discourse: Problematizing the "problem" of women and minorities in science and engineering* (Master's thesis). University of Arizona, Tucson, AZ.

Metcalf, H. (2010). Stuck in the pipeline: A critical review of STEM workforce literature. *InterActions: UCLA Journal of Education and Information Studies, 6*(2), Article 4. Retrieved from http://escholarship.org/uc/item/6zf09176

Metcalf, H. E. (2011). *Formation and representation: Critical analyses of identity, supply, and demand in science, technology, engineering, and mathematics* (doctoral dissertation). Retrieved from the University of Arizona Campus Repository, http://www.azu_etd_11521_sip1_m.pdf

Morris, J. H., & Lee, P. (2004). The incredibly shrinking pipeline is not just for women anymore. *Computing Research News, 16*(3). Retrieved from http://www.cra.org/CRN/Articles/may04/morris.lee.html

National Academy of Sciences. (2007). *Rising above the gathering storm: Energizing and employing America for a brighter economic future.* Washington, DC: The National Academies Press.

National Academy of Sciences, Committee on Science, Engineering, and Public Policy (COSEPUP). (2005). *Rising above the gathering storm: Energizing and employing America for a brighter economic future.* Washington, DC: National Academies Press.

National Center for Education Statistics. (2009). *Students who study science, technology, engineering, and mathematics (STEM) in postsecondary education, 2009.* NCES Stats in Brief. (NCES 2009-161). Washington, DC: National Institute of Education Sciences, U.S. Department of Education.

National Center for Education Statistics. (2011, April). *Postsecondary awards in science, technology, engineering, and mathematics, by State: 2001 and 2009.* Washington, DC: National Institute of Education Sciences, U.S. Department of Education.

National Governors Association. (2007). *Innovation America: Building a science, technology, engineering and math agenda.* Washington, DC: National Governors Association Center for Best Practices.

National Research Council. (1986). *Engineering infrastructure diagramming and modeling.* Washington, DC: National Academies Press.

National Science Foundation. (2012). *Computing education for the 21st century (CE21): Program solicitation.* Retrieved from http://www.nsf.gov/pubs/2012/nsf12527/nsf12527.htm

Nelson, R. (2002). *National innovation systems: A retrospective on a study in "systems of innovation: Growth, competitiveness, and employment," Vol. 2.* Northampton, MA: Elgar Publishing.

NSF Directorate for Engineering. (2010). *The role of the National Science Foundation in the innovation ecosystem.* Washington, DC: National Science Foundation.

Organisation for Economic Co-operation and Development. (2010). *The OECD innovation strategy: Getting a head start on tomorrow.* Retrieved from http://www.oecd.org/document/15/0,3343,en_2649_34273_45154895_1_1_1_1,00.html#TOC

Patton, M. (2008). *ATE projects impact.* Washington, DC: American Association of Community Colleges.

Phelps, L. A. (2012). Regionalizing postsecondary education for the twenty-first century: Promising innovations and capacity challenges. *New Directions for Community Colleges, 157,* 5–16.

Sommers, D. (2009). National labor market projections for community college students. Occupational outlook for community college students. *New Directions for Community Colleges, 146,* 33–52.

Teitelbaum, M. S. (2003). Do we need more scientists? *Public Interest, 153,* 40–53.

Weeks, P. (2009). The outlook in engineering-related technology fields. Occupational outlook for community college students. *New Directions for Community Colleges, 146,* 69–76.

World Economic Forum. (2011). *The global competitiveness report: 2011–2012.* Geneva, Switzerland: Author.

Xie, Y., & Shauman, K. (2003). *Women in science: Career processes and outcomes.* Cambridge, MA: Harvard University Press.

Zinth, K. (2006). *Recent state STEM initiatives.* Denver, CO: Education Commission of the States.

PART II

A New Dimension to the Discourse on Minority Students, STEM, and Community Colleges

4

AN EXPECTANCY-VALUE MODEL FOR THE STEM PERSISTENCE OF NINTH-GRADE, UNDERREPRESENTED MINORITY STUDENTS

Lori Andersen and Thomas J. Ward

The domestic need for STEM innovators and experts is both critical and nationally recognized (National Science Board, 2010). The proportion of U.S. students who majored in the sciences or engineering is much lower than in other countries, and 35% of the PhDs in the domestic science and engineering workforce are foreign born (Atkinson & Mayo, 2011). Meanwhile, a large amount of domestic STEM potential remains undeveloped, as evidenced by the acute underrepresentation of minorities in these disciplines. In 2008, Blacks and Hispanics comprised 31.8% of the 18- to 24-year-old U.S. population, while they represented only 15.1% of students enrolled in undergraduate engineering programs. Meanwhile, the corresponding figures for White students were 61.3% of the population and 68.1% of engineering enrollment (NSF, 2012). As ability is distributed equally among students of all races (Nisbett et al., 2012), this disproportionate representation is evidence that many Black and Hispanic students who have ability in STEM are not served well by the educational system; the undeveloped talents of these students are a loss to society.

Demographic trends in the United States indicate that population diversity is rapidly increasing. Therefore, it is important to understand the variables that facilitate STEM persistence for talented Black and Hispanic students, not only to provide equitable outcomes for these students compared to the outcomes attained by their White and Asian peers, but also to ensure the viability of the STEM workforce. Of course, these outcomes will only be attained after students take appropriate science and mathematics coursework in high school, ensuring their readiness to enter the postsecondary STEM pipeline. Therefore, achieving a greater understanding of adolescents' decisions to embark on a trajectory of STEM talent development through appropriate high school coursetaking is important to increasing the numbers of students who opt to do so.

This chapter presents the results of an investigation into the variables related to ninth-grade, underrepresented (URM) minority students' decisions to stay in or opt out of the STEM pipeline.

Framework

The theoretical framework for this study is based on the Eccles et al. (1983) expectancy value model of achievement-related choices. Based on this model, students' decisions to continue in the STEM pipeline are affected by their expectations for success in and the relative importance that they give to each available option. Expectations for success are represented by self-beliefs about ability, such as science and mathematics self-efficacy. Relative importance is described by *subjective task value* (STV), that describes the importance of taking mathematics and science courses in terms of four elements: (a) the utility value as related to the student's future goals, (b) the intrinsic value, (c) the attainment value (the consistency of mathematics and science with the student's identity), and (d) the cost, such as time taken away from other activities or the negative responses of the student's peers (Eccles, 2009). Subjective task value is individually constructed based on the processing of inputs from culture, socializers, and experiences. The Eccles et al. (1983) model has been used to explain gender differences in career choices, but it has not yet been used with samples of URM students.

This investigation focuses on the plans of ninth-grade, URM students to continue their studies of mathematics and science because previous research has shown that reentry into the STEM pipeline is rare after high school and that career plans made in high school are good predictors of future completion of STEM degrees (Maltese & Tai, 2011; Syed, Azmitia, & Cooper, 2011; Tai, Liu, Maltese, & Fan, 2006). Before high school all students stay in the science and mathematics pipeline by default. In contrast, students travel different coursework or career preparation paths in high school based on perceived ability, motivation, and opportunity. Academic preparation in high school that includes advanced mathematics is a key consideration in post-secondary studies of science and mathematics (Lee & Luykx, 2006). Therefore, increased understanding of the factors that affect plans to persist could be useful in increasing the numbers of students who pursue STEM careers.

STEM Persistence

While variables associated with STEM persistence have been studied using national data, attention has generally been focused on students who have achieved a STEM degree and the relative deficits of those students who have not done so. However, the diversity of study samples has been limited and most researchers have assumed that variables within explanatory models operate identically across all racial and socioeconomic groups because they have treated race as a single predictor variable in a model (e.g., Maltese & Tai, 2011; Mau, 2003). Extensive reviews of this literature already exist (e.g., Lee & Luykx, 2006; Maltese & Tai, 2011), space limitations preclude a detailed review of this body of research here. Briefly, previous research has identified causes of leakage of potential talent such as deficits in preparatory coursework (Lee & Luykx, 2006). Early interest has been identified as a good predictor of who will earn a STEM degree (Tai et al., 2006). Students who have taken a greater number of, and more rigorous, mathematics and science courses are more likely to pursue STEM degrees (Maltese & Tai, 2011). Fewer Black and Hispanic students completed advanced coursework compared to their Asian and White peers. Nonetheless, those who did were equally as likely to complete STEM degrees (Tyson, Lee, Borman, & Hanson, 2007). Other research has shown that students from underrepresented groups are at a greater risk of leaving a STEM major (Bonous-Harnmarth, 2000). In sum, most previous research has revealed the necessary academic paths toward a STEM degree (advanced high school mathematics and science) and the demographics of the students who are more likely to achieve STEM degrees (Asian, White, and higher socioeconomic status), but has not examined the reasons Black and Hispanic students take STEM coursework or pursue STEM degrees. One exception is a group of researchers who found that career considerations precede course-taking plans for Black high school students, placing the causal order of career choice and course taking into question (Lewis & Connell, 2005; Thompson & Lewis, 2005). Furthermore, a lack of opportunity for advanced mathematics and science coursework is a factor that affects persistence, especially for students in low-SES schools, yet deficits in preparatory coursework are often attributed to student disinterest (Thompson & Lewis, 2005). To understand how predictor variables operate in different groups necessitates analyses that are specific to that group (Lee & Luykx, 2006). This study aims to fill this gap in the literature.

Expectations for Success

Students' expectations for success in STEM are good predictors of persistence. These expectations were often operationalized as *domain-specific self-efficacy*, or one's confidence in his or her ability to successfully complete tasks within a domain. Self-efficacy was more important than achievement to occupational choice decisions (Bandura, Barbaranelli, Caprara, & Pastorelli, 2001; Eccles, 2005). Simpkins, Davis-Kean, and Eccles (2006) found that students who had higher self-efficacy or had an interest in math and science were more likely to continue studies of these subjects than their peers, and that these factors emerged above the predictive power of achievement and socioeconomic status. Mau (2003) found that mathematics self-efficacy and academic proficiency of eighth-grade students were good predictors of who would persist in aspiring to a science and engineering career. However, these studies used samples that were predominantly White. In large samples of minority middle school students, mathematics and science self-efficacy were positively related to goals and intentions for Mexican American students (Navarro, Flores, & Worthington, 2007) and for inner city, low-SES students (Fouad & Smith, 1996). Therefore, it is expected that mathematics and science self-efficacy will predict STEM persistence plans for this group of URM students.

Subjective Task Value

Two studies have been conducted using data from the National Education Longitudinal Study of 1988 (NELS: 88; National Center for Education Statistics, 2012) data that examined the effects of one or more components of STV on students' STEM persistence. First, early interest in a STEM career was found to be sufficient to sustain students in the pipeline. Specifically, Tai et al. (2006) found that even after controlling for student background and academic achievement in mathematics, students who planned on pursuing a STEM career where two to three times more likely to graduate with a college degree in the sciences than students who did not have such plans. However, the Tai et al. (2006) study lacked sufficient representation of minority students. Maltese and Tai (2011) found that eighth-grade students' perceptions of the utility of science, or STV, was a better predictor of who would complete a STEM degree than mathematics or science than were achievement test scores. These studies support the predictive usefulness of the intrinsic value and utility value components of STV, however, no previous studies have examined the four components of STV simultaneously within a nationally representative sample of URM students.

Few studies have explored racial or ethnic differences in subjective task value or occupational choice for underrepresented minority students. Zarrett and Malanchuk (2005) studied influences on Black students' decisions to pursue careers in information technology and found that these students were equally as likely to consider a career in computers as were White students. They found significant effects of students' perceived ability, value of a domain, and the influence of socializers and peers on students' decision to pursue an information technology career. Lewis and Connell (2005) found that most Black high school students' course taking decisions were based on utility value. These findings support the relevance of STV to URM students' career decisions.

Although STV is often discussed as an important predictor of academic choices, there have been no empirical studies of URM high school students' STV for STEM. According to expectancy-value theory, students who place a high STV on mathematics and science should be motivated to take such coursework. Subjective task value varies within and across racial and ethnic groups because of the differential effects of culture and socializers (Simpkins & Davis-Kean, 2005). For example, attainment value is determined by the compatibility of the task with the individual's identity, therefore other components of identity such as race, ethnicity, gender, and culture will affect the STV that individuals construct for science and mathematics (Eccles, 2009).

Race, Ethnicity, Culture, and STV

The four components of STV are each affected by the racial, ethnic, and cultural backgrounds of the individual and the interactions of the individual with STEM culture. Underrepresented minority students are rarely presented with same-race role models or prominent historical figures in science and mathematics and this may prevent these students from identifying with STEM domains because they may feel as though they must be assimilated and give up their racial identity to succeed (Cooper, 2011). Without evidence of prior successes of people from similar backgrounds, URM students may be less likely to view science and mathematics coursework as having a high utility value. However, a majority of Black students' science and mathematics course taking decisions were based on utility value or interest (Lewis & Connell, 2005). Lower subjective task values for mathematics and science coursework will reduce the likelihood that URM students will decide to persist.

Incompatible Identities

Stereotypes that are associated with STEM may conflict with components of students' identities and prevent the development of a science identity (Lewis & Connell, 2005). Science is a subculture of White, male, Western culture (Barba, 1998; Hines, 2003) that is perceived as competitive, individualistic, cut-throat, and isolated while many URM students' learning styles demonstrate preferences for collaboration, group work, cooperation, and a sense of community (Lewis & Connell, 2005; Seymour & Hewitt, 1997). These points of potential cultural conflict imply that URM students may have lower degrees of identification with, and thus a lower degree of attainment value for, science than non-URM students. Lower science identity reduces the attainment value and the subjective task value of science and is also a barrier to persistence. Differences in the subjective task value that individuals hold for science and mathematics explain differences in their STEM persistence.

This study investigated the expectations for success and the subjective task value that Black and Hispanic students have for science and mathematics through examination of the relative importance of factors such as self-efficacy, attainment value, utility value, intrinsic interest, and cost on underrepresented minority students' decisions to stay in or opt out of the STEM talent pool. Based on the Eccles et al. (1983) model, it is hypothesized that students who have high expectations for success in STEM, have an intrinsic interest in STEM, see a high degree of utility in taking science and mathematics courses related to their future goals, find science and mathematics coursework consistent with their identity, and have a positive perception of the cost of taking science and mathematics courses are more likely to enroll in these courses, be prepared for college-level STEM studies, and complete STEM degrees. The current investigation explores the relative importance of these factors that underlie URM students' decisions to continue in or to opt out of a STEM education trajectory. Based on the Eccles et al. (1983) model, the following hypothesis was made: there will be direct effects for gender, socioeconomic status, students expectations for success (mathematics and science self-efficacy), and subjective task value (STEM utility value, mathematics and science intrinsic value, mathematics and science attainment value, and cost) on ninth-grade, URM students decisions to persist in STEM.

Method

The High School Longitudinal Study of 2009 (Ingels et al., 2011) is a secondary longitudinal study from the National Center for Education Statistics (NCES, 2011). The data used in this study come from the base year, when the students were in the fall of the ninth grade. The sample is representative of ninth-grade students in public and private schools in the United States in 2009 (N = 21,144). For this analysis, the sample was reduced to the 5,733 students who self-identified as Black (37.8%) or Hispanic (62.2%). The sample was 49.1% male and 50.9% female.

STEM Persistence Status

In order to determine students' STEM persistence status, students were asked to identify the occupation they expected to have at age 30. Students who selected: computer and mathematical; architecture and engineering; life, physical, and social sciences; or healthcare practitioners and technical occupations were identified as having elected to persist. An alternate criterion was devised because a large number of students (28.2%) responded with "don't know." If a student planned on taking 4 years of mathematics, 4 years of science, and at least one Advanced Placement or International Baccalaureate mathematics or science course during high school, the student was included. Students who met either of the two criteria—identification of a future STEM occupation or indication of intent to persist—were assigned the dependent variable value of "planned to persist."

High-Ability Criteria

High-ability was operationalized by selecting students who scored in: (a) the top 10% of the mathematics achievement test or (b) in the top 10% of

TABLE 4.1 High Ability Criteria by Race

	Black	Hispanic
Mathematics Achievement Score	49.59	51.56
Mathematics and Science GPA	4.0	4.0

Source: High School Longitudinal Study of 2009. Tabulations by Author. Values not weighted.

eighth-grade mathematics and science GPA based on a definition of giftedness (NAGC, 2011). In alignment with recommendations for the identification of ability in underrepresented groups (e.g., Lohman, 2005), group-specific norms were used (Table 4.1). Students who scored in the top 10% within their racial group (Black or Hispanic) were identified. The GPA criterion was selected because previous performance in a domain is a good predictor of future performance (Cross & Coleman, 2005). Students who exhibited high ability through one of the two indicators were considered to have had high ability.

Model

A progression of models were created to assess the effect of predictor variables including demographic variables, domain specific self-efficacies, utility value, intrinsic interest, attainment value, and perception of cost. These logistic regression models examined the effects of independent variables on one dichotomous, dependent variable (Hosmer & Lemeshow, 2000). Analyses were conducted for the URM student sample with STEM pipeline status as the dependent variable. Building the regression model in this manner allowed the examination of how the relationships between significant variables and the dependent variable evolved as additional factors were added. This process determined whether the addition of variable significantly improved the model.

Results

Preliminary gender differences in students' persistence plans were examined using chi square analysis (Figure 4.1). A gender difference was found favoring females (χ^2 = 225.21, p = .000). Females were more likely to plan to persist in STEM (Figure 4.1). Ability-based differences in URM students' persistence plans were also examined. A significant difference was found favoring high-ability students (χ^2 = 684.8, p = .000). Students who had high ability in mathematics and science were more likely to plan to persist. However, even among URM students with high ability there was substantial number of non-persisters; 45% of the high-ability URM students in this sample did not plan to persist.

A significant difference was found in persistence plans among students with higher SES (χ^2 = 2073.5, p = .000). Students in the lowest quintile of SES were

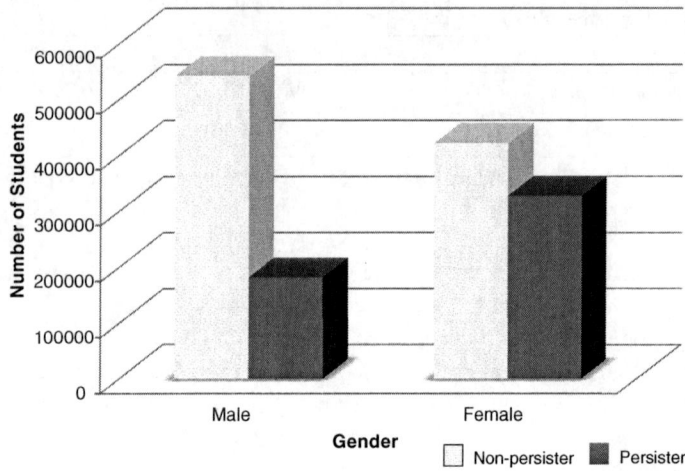

FIGURE 4.1 Persistence status of ninth-grade underrepresented minority students.

significantly less likely to plan to persist. Hierarchical logistic regression analyses were used to examine ninth-grade demographic predictors (SES, gender, and high-ability status) that previous research has identified as predictive of STEM persistence. The next step involved regressing individual expectations for success (math and science self-efficacy). Last, the subjective task value factors were added to see if these mediated the effects of SES, gender, high-ability

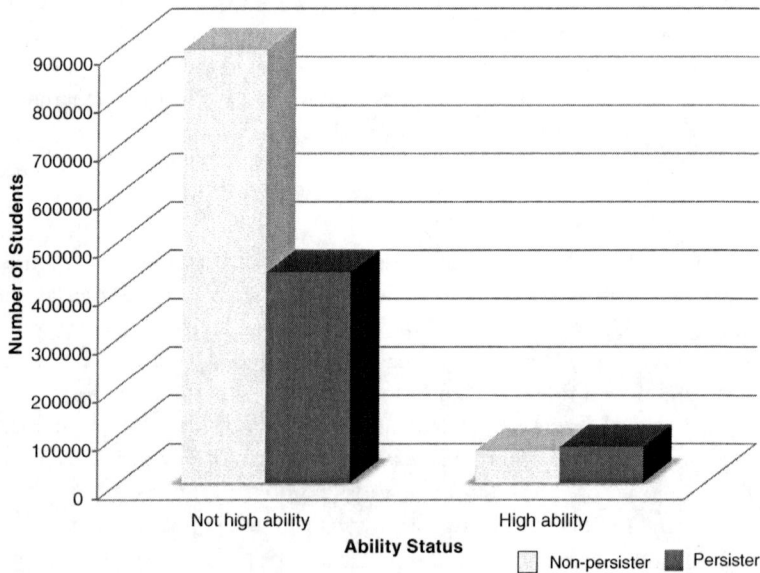

FIGURE 4.2 Persistence status of ninth-grade underrepresented minority students.

status, and self-efficacy. The descriptive statistics for the predictor variables are presented in the Appendix.

SES, Gender, and High Ability

In the first model, socioeconomic status significantly predicted planned STEM persistence; students from higher SES households were significantly more likely to plan to persist ($\chi^2 = 41.67$, $p = .000$). The addition of gender to the model significantly improved the model fit. Compared to males, females were significantly more likely to plan to persist. According to the model, females were more than twice as likely to plan to persist in STEM as males. High-ability status was also a significant predictor, students who had high ability were significantly more likely to plan to persist in STEM. According to the model, students with high ability were more than twice as likely to plan to persist.

Domain-Specific Self-Efficacy

When the individual expectations of success variables, science and math self-efficacy, were added into the model in step four, science self-efficacy was a significant predictor of planned persistence but math self-efficacy was not a

TABLE 4.2 Nested Models for the Planned STEM Persistence of Underrepresented Minority Ninth-Grade Students (N = 5,733)

Variables	1	2	3	4	Final
SES	1.648***	1.648***	1.516***	1.482***	1.405***
Gender – Female		2.319***	2.266***	2.423***	2.400***
High-Ability Status			2.125***	1.839***	1.346
Science Self-Efficacy				1.497***	1.080
Math attainment					1.163*
Science attainment					1.171*
Science intrinsic					1.373***
STEM utility					1.433***
χ^2	41.67	103.67	134.03	172.96	269.27
$\Delta\chi^2$		62.00	30.35	38.90	96.30
Df	1	2	3	4	8
Δdf		1	1	1	4
Pseudo R2	.035	.085	.106	.138	.209
ΔPseudo R2		.050	.021	.032	.071

Source: High School Longitudinal Study of 2009. Tabulations by Author. Data are weighted by W1Student.
*** $p < .001$, * $p < .05$

significant predictor. The addition of science self-efficacy significantly improved the model fit.

Subjective Task Value

The fifth step represented a significant model improvement. Students who held a higher attainment value for mathematics and science, a higher intrinsic value of science, and a higher utility value of science were more likely to plan to persist. Intrinsic math value and the two cost variables were not significant predictors of persistence.

Final Model

The effects of SES and gender were mediated somewhat by the other predictors that were added to the model but remained significant nonetheless. The effect of high ability was mediated substantially by the subjective task value variables and the effect of science self-efficacy was not significant after the more proximal expectancy value model predictors were added. Therefore, the influence of a ninth-grade, URM student's science self-efficacy was accounted for by the subjective task value predictors entered in step five. The final model explained 20.9% of the variation in plans to persist.

These findings suggest that ninth-grade, URM students who have higher SES, are female, have a higher attainment value, have a higher intrinsic value of science, and higher STEM utility value were more likely to plan to persist in STEM. The majority of ninth-grade, URM students who planned to persist were female (n = 1,227, 60.7%). The final model indicated that females were 2.4 times more likely than males to plan to persist. These gender differences remained strong predictors even after expectations for success and subjective task value variables were accounted for. Contrary to observed gender differences in the current population of STEM undergraduate students (NSF, 2012), these findings suggest that ninth-grade, URM females are more likely to plan to persist in STEM than males. Furthermore, the findings also show that SES is a significant predictor of persistence. Interestingly, once all other predictors were considered, mathematics and science self-efficacy did not appear to play a significant role in persistence plans. This finding is contrary to previous research that showed significant effects of self-efficacy on goal and intentions in URM eighth-grade students (Fouad & Smith, 1996; Navarro et al., 2007).

Discussion

The goal of this chapter was to examine the dynamic process by which ninth-grade, URM students make STEM persistence plans. It was hypothesized that self-efficacy and subjective task value factors would mediate the effects of gender and SES on STEM persistence plans and that these non-demographic variables would be significantly and positively related to persistence. The findings of this study partially support the hypothesis. Of the nine non-demographic variables examined through logistic modeling, four were found to be significantly and positively related to these students' persistence plans: math attainment value, science attainment value, science intrinsic value, and STEM utility value. The other five variables: science self-efficacy, mathematics self-efficacy, cost-time, cost-peers, and math intrinsic value were not significant predictors of persistence. The variables of SES and gender remained significant in the final model. Therefore, the findings of this study support that URM students' persistence plans are shaped by their subjective task values, specifically their math and science attainment values, science intrinsic values, and STEM utility values.

One explanation for why mathematics and science self-efficacy were not significant predictors of persistence plans in this sample may be differences in URM students' perceptions of locus of control and barriers to opportunities compared to those of White students. Expectations for success may not be adequately represented by domain-specific efficacy for URM students who may perceive that they have limited career opportunities in science despite their abilities, perhaps due to the history of underrepresentation of Blacks and Hispanics in the sciences (Lewis, 2003; Lynch, 2011). Underrepresented minority students who have high ability are likely to make their career decisions using additional information such as prevailing stereotypes and perceptions of job opportunities. Another explanation could be the disproportionate effects of poverty on URM students. A young Black or Hispanic student is eight times as likely to be in a high-poverty school compared to a White student (U.S. Department of Education, 2011) and is likely to have had fewer educational experiences to prepare for high school science. The compounded effects of poverty and institutionalized racism may negate the effect of self-efficacy on persistence plans (Graham & Hudley, 2005).

The logistic regression analyses provided a ranking of the relative importance of the four components of STV. The results of this study support the idea that STEM utility value and science attainment value are better predictors of the persistence plans of ninth-grade, URM students than perceptions of cost or intrinsic value. This finding compares favorably to the work of Oyserman and Destin (2010) who explain differences in the academic attainment of URM students as related to preferences for identity-congruent actions over identity-incongruent actions. Students who believe that science and mathematics are identity-congruent will have a high attainment value for these courses. The predictive value of attainment value over intrinsic value in this study can be explained by the theory of identity-based motivation. Aschbacher, Li, and Roth (2009) documented strong relationships between aspirations, persistence, and identity in their longitudinal study of a diverse sample of high school students. Eccles (2009) has reframed her expectancy value model using identity as a basis for motivation. Thus, it seems for many URM students their own identity is incongruent with their perception of a STEM identity and this is a barrier to persistence.

Implications

These findings have important implications for interventions that could facilitate students' development of subjective task value for science and mathematics and subsequent persistence in STEM. Attainment value is directly related to the degree to which students identify with mathematics and science, while utility value is determined by the connections between STEM coursework and students' personal goals. Interventions should address factors such as providing: opportunities to explore STEM identities, enrichment experiences that allow students to develop an interest in and enjoyment of science, culturally responsive science instruction, and accurate information about STEM careers. Community colleges are in a unique position to influence the career aspirations of URM students because the majority of these students enter college through community college (Fry, 2010). In this section, methods that community colleges can employ to increase URM students' subjective task value of, and future participation in, STEM are presented.

Community colleges can help URM students to decrease the incongruence between their own identities and STEM identities. Many students lack accurate knowledge regarding careers and the actual work done by engineers and scientists which leads these students to believe that the constraints of a STEM occupation are a barrier to other life goals pertaining to community, family, and leisure activities (Lewis & Connell, 2005; Taningco, Mathew, & Pachon, 2008). Current information and advice about career options is critical for students to make good decisions. Community colleges should sponsor job fairs where students can get accurate information about STEM careers and education requirements for those positions. Partnerships between college and businesses should be created that allow students to have cooperative work experiences, build confidence in their abilities, and make connections with people who work in STEM. Through positive mentoring and cooperative work experiences students will be able to see beyond the stereotypes and view a STEM career as a potentially lucrative, satisfying lifestyle that is compatible with their racial, social, and cultural identities.

The way that courses are taught is important to the recruitment and retention of URM students in the STEM disciplines. First, introductory science courses need a change of purpose, from the traditional view of "weeding out" those who may be believed less capable to a more progressive view of inspiring interest, scaffolding learning for all students, and scouting for talent. To serve this new purpose will require using research-based principles of teaching and learning with established effectiveness that are culturally responsive. Students must learn about the nature of science and how science knowledge is created so that they can realize that their own ideas are valuable. Teaching strategies that emphasize active learning and collaboration such as problem-based learning or inquiry learning are culturally responsive because students can investigate issues that are relevant to them and participate in building scientific knowledge. Introductory courses must be interesting and engaging to inspire students to continue studying that discipline (Hrabowski, 2011). If students find course content to not be relevant or interesting, students are likely to switch to a program that is relevant and interesting. Introductory courses must be reinvented to prevent a URM student's first course in a STEM discipline from becoming their last. Introductory level courses instructors should strive to inspire student interest in their subjects and to engage all students through culturally responsive teaching practices.

Minority students may consider science as foreign because they do not learn about any scientists or inventors from backgrounds similar to their own or encounter scientists in their communities (Hines, 2003). These students may internalize the idea that they cannot perform science or may feel that they must lose their racial identity to be assimilated. Culturally responsive teaching methods can increase URM student interest in science courses and facilitate students' crossings between their own culture and the culture of science (Hines, 2003). Barba (1998) describes how science teaching must be more harmonious with culturally syntonic variables. For example, science classes that emphasize individual competition and where grading is on a curve do not fit well with the learning styles of culturally diverse students who prefer to work more collaboratively and to develop extended networks of support among their peers. Science instructors should reduce language barriers to learning by connecting science language and students' native languages to develop students' skills in making "border crossings" between the different worlds and identities they navigate in life (Cooper, 2011). Through these and other culturally responsive teaching practices, a STEM identity could become more congruent with students' identities and science attainment value may increase.

It is important to encourage interest in STEM because students who select careers for reasons other than interest are less likely to persist (Seymour & Hewitt, 1997). To build interest, community colleges should sponsor STEM clubs and competitions and actively recruit for these activities. The accomplishments of these participants should be recognized and lauded to create a prestige associated with belonging to these groups.

In this study, a model for the persistence plans for a group of ninth-grade, URM students was developed; however, it remains to be discovered how models for persistence may compare for different groups of ninth-grade students. Differences in the predictive models between race groups may reveal different relationships among the predictor variables due to the relative congruence or incongruence of students' identities with a STEM identity. Understanding these differences between groups of students may help educators to become more culturally responsive. The findings of this study provide a focus for interventions that could increase STEM persistence of URM students. Interventions must be grounded in research and designed with a specific purpose in mind. This analysis directs attention to the factors of STEM utility, and science attainment value, in particular. Improving the identity congruence between students' identities and STEM identities will improve the utility and attainment values students have for STEM and potentially increase the number of students who persist.

APPENDIX A Descriptive Statistics for Predictor Variables as a Function of STEM Pipeline Status for Underrepresented Minority Students

	Persisters (N = 2,020)	Non-persisters (N = 3,713)	Total (N = 5,733)
	M (SD)	M (SD)	M (SD)
High-Ability Status[a]	648	530	1,178
Female[a]	1,227	1,593	2,913
Male[a]	793	2,120	2,820
SES[b]	−.29 (.70)	−.51 (.64)	−.53 (.67)
Math Self-Efficacy[b]	.07 (.95)	−.11 (.92)	−.05 (.93)
Science Self-Efficacy[b]	.11 (.85)	−.21 (.88)	−.10 (.88)
STEM Utility Value[b]	.13 (.95)	−.39 (.82)	−.21 (.90)
Science Intrinsic Value[b]	.09 (1.05)	−.31 (.72)	−.17 (.87)
Math Intrinsic Value[b]	.08 (1.04)	−.20 (.82)	−.10 (.91)
Cost-Peers[b]	.07 (1.03)	−.12 (.98)	−.05 (1.00)
Cost-Time[b]	.15 (.95)	.00 (1.00)	.05 (.98)
Math Attainment Value[b]	.15 (.99)	−.15 (.97)	−.05 (.99)
Science Attainment Value[b]	.18 (.98)	−.31 (.90)	−.14 (.96)

Source: High School Longitudinal Study of 2009. Tabulations by Author. Data are weighted by W1Student. Note: [a]Frequency. [b]Standardized score for the population of HSLS: 2009 with an approximate mean of zero and approximate standard deviation of one.

References

Aschbacher, P. R., Li, E., & Roth, E. J. (2009). Is science me? High school students' identities, participation and aspirations in science, engineering, and medicine. *Journal of Research in Science Teaching, 47*(5). doi:10.1002/tea.20353

Atkinson, R. D., & Mayo, M. (2011). *Refueling the U.S. innovation economy: Fresh approaches to science, technology, engineering and mathematics (STEM) education.* Washington, DC: The Information Technology and Innovation Foundation.

Bandura, A., Barbaranelli, C., Caprara, G. V., & Pastorelli, C. (2001). Self-efficacy beliefs as shapers of children's aspirations and career trajectories. *Child Development, 72,* 187–206. doi:10.1111/1467-8624.00273

Barba, R. H. (1998). *Science in the multicultural classroom: A guide to teaching and learning* (2nd ed.). Boston, MA: Allyn and Bacon.

Bonous-Harnmarth, M. (2000). Pathways to success: Affirming opportunities for science, mathematics, and engineering majors. *Journal of Negro Education, 69,* 92–111.

Cooper, C. R. (2011). *Bridging multiple worlds: Cultures, identities, and pathways to college.* New York, NY: Oxford University Press.

Cross, T. L., & Coleman, L. J. (2005). School-based conception of giftedness. In R. J. Sternberg & J. E. Davidson (Eds.), *Conceptions of giftedness* (2nd ed., pp. 52-62). Cambridge, UK: Cambridge University Press.

Eccles, J. S. (2005). Studying gender and ethnic differences in participation in math, physical science, and information technology. *New Directions for Child and Adolescent Development, 110,* 7–14. doi:10.1002/cd.146

Eccles, J. S. (2009). Who am I and what am I going to do with my life? Personal and collective identities as motivators of action. *Educational Psychologist, 44,* 78–89. doi:10.1080/00461520902832368

Eccles, P. [Parsons], J. S., Adler, T. F., Futterman, R., Goff, S. B., Kaczala, C. M., Meece, J. L., & Migley, C. (1983). Expectations, values and academic behaviors. In J. T. Spence (Ed.), *Perspectives on achievement and achievement motivation* (pp. 75–146). San Francisco, CA: W. H. Freeman.

Fouad, N. A., & Smith, P. L. (1996). A test of a social cognitive model for middle school students: Math and science. *Journal of Counseling Psychology, 43,* 338–346. doi:10.1037/0022-0167.43.3.338

Fry, R. (2010). *Minorities and the recession-era college enrollment boom.* Washington, DC: Pew Research Center.

Graham, S., & Hudley, C. (2005). Race and ethnicity in the study of motivation and competence. In A. J. Elliot & C. S. Dweck (Eds.), *Handbook of competence and motivation* (pp. 392–413). New York, NY: Guilford.

Hines, S. M. (2003). *Multicultural science education: Theory, practice, and promise.* New York, NY: Peter Lang.

Hosmer, D. W., & Lemeshow, S. (2000). *Applied logistic regression* (2nd ed.). New York, NY: Wiley.

Hrabowski, F. A. (2011). Boosting minorities in science. *Science, 331*(6014), 125. doi:10.1126/science.1202388

Ingels, S. J., Pratt, D. J., Herget, D. R., Burns, L. J., Dever, J. A., Ottem, R., ... Leinwand, S. (2011). *High School Longitudinal Study of 2009 (HSLS:09). Base-year data file documentation* (NCES 2011-328). Retrieved from http://nces.ed.gov/surveys/hsls09/hsls09_data.asp

Lee, O., & Luykx, A. (2006). *Science education and student diversity: Synthesis and research agenda.* New York, NY: Cambridge University Press.

Lewis, B. F. (2003). A critique of literature on the underrepresentation of African Americans in science: Directions for future research. *Journal of Women and Minorities in Science and Engineering, 9,* 361–373. doi:10.1615/JWomenMinorScienEng.v9.i34.100

Lewis, B. F., & Connell, S. (2005). African American students' career considerations and reasons for enrolling in advanced science courses. *The Negro Educational Review, 56,* 221–232.

Lohman, D. F. (2005). The role of nonverbal ability tests in identifying academically gifted students: An aptitude perspective. *Gifted Child Quarterly, 49*(2), 111–138. doi:10.1177/001698620504900203

Lynch, S. J. (2011). Equity and U.S. science education policy from the G.I. Bill to NCLB: From opportunity denied to mandated outcomes. In G. E. DeBoer (Ed.), *The role of public policy in K-12 science education* (pp. 305–354). Charlotte, NC: Information Age.

Maltese, A. V., & Tai, R. H. (2011). Pipeline persistence: Examining the association of educational experiences with earned degrees in STEM among U.S. students. *Science Education, 95,* 877–907. doi:10.1002/sce.20441

Mau, W. C. (2003). Factors that influence persistence in science and engineering career aspirations. *The Career Development Quarterly, 51,* 234–243. doi:10.1002/j.2161-0045.2003.tb00604.x

NAGC. (2011). *Redefining giftedness for a new century: Shifting the paradigm.* Retrieved from http://www.nagc.org/index2.aspx?id=6404

National Center for Education Statistics. (2011). *High school longitudinal study: 2009 EDAT extract codebook.* Retrieved from http://nces.ed.gov/surveys/hsls09/hsls09_data.asp

National Center for Education Statistics. (2012). *National education longitudinal study of 1988 (NELS).* Retrieved from http://nces.ed.gov/surveys/nels88/

National Science Board. (2010). *Preparing the next generation of STEM innovators: Identifying and developing our nation's human capital.* Retrieved from http://www.nsf.gov/nsb/publications/2010/nsb1033.pdf

Navarro, R. L., Flores, L. Y., & Worthington, R. L. (2007). Mexican American middle school students' goal intentions in mathematics and science: A test of social cognitive career theory. *Journal of Counseling Psychology, 54,* 320–335. doi:10.1037/0022-0167-54.3.320

National Science Foundation. (2012, June). *Women, minorities, and persons with disabilities in science and engineering.* Retrieved from http://www.nsf.gov/statistics/wmpd/race.cfm

Nisbett, R. E., Aronson, J., Blair, C., Dickens, W., Flynn, J., Halpern, D. F., & Turkheimer, E. (2012). Intelligence: New findings and theoretical developments. *American Psychologist.* doi:10.1037/a0026699

Oyserman, D., & Destin, M. (2010). Identity-based motivation: Implications for intervention. *The Counseling Psychologist, 38,* 1001–1043. doi:10.1177/0011000010374775

Seymour, E., & Hewitt, N. M. (1997). *Talking about leaving: Why undergraduates leave the sciences.* Boulder, CO: Westview Press.

Simpkins, S. D., & Davis-Kean, P. E. (2005). The intersection between self-concepts and values: Links between beliefs and choices in high school. *New Directions for Child and Adolescent Development, 110,* 31–47. doi:10.1002/cd.148

Simpkins, S. D., Davis-Kean, P. E., & Eccles, J. S. (2006). Math and science motivation: A longitudinal examination of the links between choices and beliefs. *Developmental Psychology, 42,* 70–83. doi:10.1037/0012-1649.42.1.70

Syed, M., Azmitia, M., & Cooper, C. R. (2011). Identity and academic success among underrepresented ethnic minorities: An interdisciplinary review and integration. *Journal of Social Issues, 67*(3), 442–468.

Tai, R. H., Liu, C. Q., Maltese, A. V., & Fan, X. (2006). Planning early for careers in science. *Science, 312,* 1143–1144. doi:10.1126/science.1128690

Taningco, M. T. V., Mathew, A. B., & Pachon, H. P. (2008). *STEM Professions: Opportunities and challenges for Latinos in science, technology, engineering, and mathematics.* Los Angeles, CA: The Tomas Rivera Policy Institute. Retrieved from http://www.trpi.org/PDFs/STEM%20Lit_Final.pdf

Thompson, L. R., & Lewis, B. F. (2005, April/May). Shooting for the stars: A case study of the mathematics achievement and career attainment of an African American male high school student. *The High School Journal,* 6–18.

Tyson, W., Lee, R., Borman, K. M., & Hanson, M. A. (2007). Science, technology, engineering, and mathematics (STEM) pathways: High school science and math coursework and postsecondary degree attainment. *Journal of Education for Students Placed at Risk, 12,* 243–270. doi:10.1080/10824660701601266

U. S. Department of Education, National Center for Education Statistics. (2011). *The condition of education 2011,* NCES 2011-033. Washington, DC: U.S. Government Printing Office.

Zarrett, N. R., & Malanchuk, O. (2005). Who's computing? Gender and race differences in young adults' decisions to pursue an information technology career. *New Directions for Child and Adolescent Development, 110,* 65–84. doi:10.1002/cd.150

5

THE EFFECT OF NON-COGNITIVE PREDICTORS ON ACADEMIC INTEGRATION MEASURES

A Multinomial Analysis of STEM Students of Color in the Community College

Marissa Vasquez Urias, Royel M. Johnson, and J. Luke Wood

The need to engage more students from underrepresented populations in science, technology, engineering, and mathematics (STEM) education is a topic that has gained increasing attention in recent years. Recognizing that America could lose its competitive edge in the global market economy in STEM fields, the federal government (by way of the National Science Foundation) has made a concerted effort to focus on enhancing the proportion of women and underrepresented minorities in the STEM pipeline (U.S. GAO, 2005). Given that the community college serves as the overwhelming pathway for historically underrepresented and underserved students into higher education (Nevarez & Wood, 2010), the institution serves as a logical focal point for programs, policies, monies, and research designed to enhance STEM degree production. With respect to the latter, community college researchers have begun to respond to the clarion call to better understand and (as a result) better facilitate outcomes (e.g., persistence, attainment, achievement, transfer) for STEM students. Most notably, in 2010, the *Journal of Women and Minorities in Science and Engineering* featured a special issue on STEM and community colleges that was guest edited by Soko Starobin, Frankie Laanan, and Carol Burger. Among other contributions, this special issue did an exemplary job on several fronts: (a) articulating the critical role that community colleges can (and do) play in the STEM pipeline (Starobin & Laanan, 2010); (b) providing a historical perspective on STEM production in community colleges (Hardy & Katsinas, 2010); and (c) discussing the effect of developmental education on the STEM pipeline (Hagedorn & DuBray, 2010). All the while, the authors delineated disparities by gender and ethnicity (e.g., Lester, 2010).

Indeed, such disparities are glaring. Data from the 2009 Beginning Postsecondary Students Longitudinal Study (BPS), a national survey of students who began their college education during the 2003–04 school year and were tracked over a 6-year time frame, illustrate success differences. In this sample, 8.8% of White students began their collegiate careers as science, technology, engineering, or math majors. While Black and Asian students had slightly higher enrollments in STEM (at 9.3% and 13.9%, respectively), Hispanic/Latino students had slightly lower enrollments (at 7.7%) (BPS, 2009a). However, data on students who remained in STEM fields illustrates a different pattern. While 33.7% of White and 44.0% of Asian students remained in STEM fields, only 25.5% of Black and 17.8% of Hispanic/Latino students did so. In contrast, 36.1% of Blacks and 46.2% of Hispanic/Latino students changed their major to pursue non-STEM fields. These percentages are particularly concerning (especially for Hispanic/Latino students) in light of the lower White (32.8%) and Asian (28.1%) field change data (see Table 5.1). However, staying in a major does not necessarily indicate success. Here too, disparities are evident. While 41.5% of White and 47.6% of Asian students either attained a certificate/degree or transferred to a four-year institution during this time frame, fewer percentages of Black (31.8%) and Hispanic/Latino (30.2%) students did so during the same period (BPS, 2009b). The picture painted by these data illustrates that, while enrollments in STEM majors are nearly the same (with Blacks being slightly higher and Hispanic/Latinos being slightly lower than their White counterparts), Black and Latino students are more likely to leave the STEM fields and less likely to attain a degree or transfer than their White and Asian counterparts. As an aside, it should be noted that data for Asian students are somewhat limited as a point of comparison, given numerous within group differences that are shielded when viewed under one racial umbrella. For example, when disaggregated by ethnic group, prior research has shown that southeast Asian students (e.g., Hmong, Laotian, Vietnamese, Cambodian) face socioeconomic, cultural, and other challenges that inhibit their success in education and complicate model minority stereotypes (Ngo, 2006; Ngo & Lee, 2007).

Given evident disparities, more research is needed to better understand what factors propel historically underrepresented and underserved students (specifically students of color) to succeed in STEM. Based upon prior research on students of color in community colleges, the authors perceived that non-cognitive variables would play an important role in whether these students succeeded in STEM contexts. Non-cognitive factors "capture students' perceptions of their educational experiences and the corresponding affective responses (e.g.,

TABLE 5.1 Field Changes to Non-STEM Majors

	White	Black	Hispanic/Latino	Asian	All
Changed to non-STEM field	32.8%	36.1%	46.2%	28.1%	35.1%
Stayed in STEM field	33.7%	25.5%	17.8%	44.0%	31.0%
Left PSE with no degree	33.5%	38.4%	36.0%	27.9%	33.9%

feelings, emotions)" (CCSM, 2012, para. 8). While numerous non-cognitive variables are explored within educational contexts, much of the literature on persistence and achievement investigates the effects of self-efficacy, locus of control, degree utility, focus/effort (sometimes referred to as action control) on student outcomes.

Self-efficacy emanates from the work of Albert Bandura. Bandura (1977, 1986) defined self-efficacy as confidence in one's ability to control their emotions, behaviors, and actions for the purpose of obtaining desired outcomes. In academic contexts, self-efficacy is often employed as a measure of a student's confidence in their academic abilities (Torres & Solberg, 2001; Solberg, O'Brien, Villarreal, Kennel, & Davis, 1993). *Locus of control* is a term used to describe students who lack a personal sense of control of their behaviors and lives; rather they attribute the control of their lives to forces outside of their personal power (Faison, 1993). In general, locus of control is viewed on a continuum from external locus to internal locus. The term *locus* indicates the location where *control* is situated; for example, students who have external locus believe they lack control over their lives, while those with an internal locus believe they control their own fate. In education, control refers to their academic trajectories, experiences, and outcomes. Generally, students with an internal locus are more likely to persist and succeed than those with an external locus (Bean, 2005). Another important concept explored in the literature is that of degree utility. Degree utility refers to students' perceptions of the worthwhileness of their collegiate endeavors (Bean & Metzner, 1985; Mason, 1998). Thus, utility is an evaluative perception of the usefulness of college in comparison to other potential pathways or opportunities. Often, this assessment is made in light of prior academic experiences which can serve to positively or negatively reify the value of school (Wood, 2011). Action control "refers to a student's ability to regulate behavior, put forth a sustained level of effort, and persist in the face of difficulties" (Barber, 2011, p. 37). Wood and Palmer (in press) have discussed this concept using the term *focus*, which they discuss as a student's level of directed effort towards their academic endeavors. In essence, greater directedness is associated with positive academic outcomes.

Extensive research has illustrated that non-cognitive variables are significant predictors of student outcomes (e.g., persistence, achievement, attainment) (Aguayo, Herman, Ojeda, & Flores, 2011; Bong, 2001; Choi, 2005; Gore, 2006; Majer, 2009; Pajares & Schunk, 2001; Vuong, Brown-Welty, & Tracz, 2010; Zimmerman, 2000). While the direct effects of these variables on academic outcomes are documented (Abd-El-Fattah, 2005; Brown et al., 2008), other research has shown their effect on outcomes through indirect mechanisms. Take self-efficacy as an example; prior research has shown that self-efficacy reduces student stress and anxiety (Abd-El-Fattah, 2005; Solberg & Villarreal, 1997; Torres & Solberg, 2001; Zajacova, Lynch, & Espenshade, 2005), facilitates smoother adjustments to college environments (Chemers, Hu,

& Garcia, 2001; Ramos-Sánchez & Nichols, 2007), fosters higher levels of college satisfaction (DeWitz & Walsh, 2002), and aids students with identifying challenging goals (Brown et al., 2008).

More importantly (at least in context of this current research), non-cognitive variables are also found to be related to important measures of academic incorporation (see Kane, Beals, Valeau, & Johnson, 2004; Lenaburg, Aguirre, Goodchild, & Kuhn, 2012; Mesa, 2012; Thompson, 2001). For example, Hirschy, Bremer, and Castellano (2011) propagated a model of community college student success in career and technical education. Their model suggested that motivation, self-efficacy, locus of control, and coping strategies are directly related to important persistence predictors, including faculty–student interaction, student–student interaction, active learning, and collaborative learning experiences. This focus of the model mirrors that of Bean and Eaton's (2001), *psychological model of college student retention*. In this model, three types of non-cognitive outcomes (e.g., self-efficacy, internal locus of control, and coping) are identified as having a direct effect on persistence. They are also found to have an indirect effect on persistence by facilitating students' academic incorporation into campus settings.

The notion that non-cognitive predictors facilitate positive academic behaviors was the impetus for this current study. Bearing this in mind, this chapter reports on a study that examined the relationship between non-cognitive outcomes and academic integration measures for STEM students of color in the community college. In line with prior research, the researchers postulated that non-cognitive variables facilitated healthy academic patterns (e.g., interactions, service usage, time maximization) that resulted in more positive outcomes. However, departing from the current literature, this study had a more nuanced aim. Our interest was on the differential effects of non-cognitive predictors on differing types of academic patterns. Specifically, the goal of this study was to understand the relationship (if any) of non-cognitive variables to distinctive measures of academic integration, including: faculty–student interactions, meeting with academic advisors, studying at the library, and using the internet to access school library resources.

Historically, these academic patterns have been conceptualized through the lens of academic integration. As proffered by Tinto (1975, 1988, 1993), academic integration refers to academic experiences and relationships that foster students' incorporation into the academic milieu of a campus. In essence, Tinto's work suggests that greater levels of incorporation lead to greater levels of commitment to an institution and to students' academic endeavors (referred to as goal commitment). Overall, goal commitment is a positive predictor of persistence and achievement. It should be noted that Tinto's work also discusses the importance of incorporation in campus social settings (e.g., establishing campus friendships, participating in clubs and sports). This is an important distinction between this study and other research. This research focuses solely on academic integration measures as opposed to social integration. Wood (2012) has shown that academic integration tends to have more salience for community college students than social integration. In some cases, his research notes that some social integration measures can be negatively predictive (albeit slightly) of persistence.

There are two primary reasons why this study has important implications for STEM education in community colleges. First, prior research has shown that Tinto's framework (particularly the concept of academic integration) has direct applicability for students of color in STEM. Investigations of successful STEM programs serving students of color extol the importance of academic integration on student success. For example, Palmer, Davis, and Peters (2008) examined a STEM program for Black students using Tinto's work as an analytic lens. They found that the program provided a supportive environment with high expectations that fostered faculty–student interactions, peer mentorship, student support groups, and use of student services (e.g., tutoring, advising). They suggested that positive integration experiences were a primary explanation for positive student outcomes. Similarly, Palmer, Davis, and Thompson (2010) examined successful STEM initiatives focused on Black student populations. They noted that academic integration experiences through faculty–student interactions, engagement, and mentorship were successful practices among these efforts. These findings are affirmed by other research (see Flowers, 2012; Fries-Britt, Burt, & Franklin, 2012; Essien-Wood & Wood, 2013) which indicates that academic integration experiences (particularly those with faculty) are effective strategies for STEM student success.

Second, this study departs from prior research literature on community colleges by examining a wide array of non-cognitive variables. Most psychosocial research on STEM in the community colleges delimits non-cognitive explorations to general academic self-efficacy. However, self-efficacy can be manifested in a myriad of areas. This study employs a more comprehensive approach by examining two specific types of academic confidence, including math and English self-efficacy. Further, this study also examines other non-cognitive predictors such as locus of control, degree utility, and action control (also referred to as focus or effort in the research literature; see Wood & Palmer, in press). The wide range of predictors employed in this study provides for one of the most in-depth examinations of non-cognitive variables on community college success in STEM. Guided by these notions, the next section will explore the methods employed in this study.

Methods

Data employed in this chapter was derived from the Educational Longitudinal Study (hereafter referred to as ELS: 2002/2006). In general, the survey focuses on students' educational experiences, academic outcomes, as well as their personal and academic goals. ELS employs a multi-stage sampling technique, where data are collected from study participants over the course of three waves (Ingels, Pratt, Rogers, Siegel, & Stutts, 2004). Wave 1, collected in 2002, serves as the base year for the study. In this wave, data is collected from students during their sophomore year in high school (tenth grade). This wave collected data on students' perceptions and experiences in school as well as background information on respondents, their parents, their families, and friends. Two years after the initial collection, students were surveyed again. During this collection, most were seniors in high school, though a smaller demographic had dropped out, earned GEDs, or graduated early. This collection featured a wide array of variables including information on academic habits, psychosocial outcomes, personal goals, school involvement, and academic outcomes. In 2006, the most recurrent wave of ELS data was collected to better understand respondents' initial post-high school experiences in college and/or the workforce. This included data on high school outcomes, college academic and social experiences, as well as labor market earnings and satisfaction. Data in this chapter is derived from all the aforementioned waves. The dataset was delimited to students of color, who were declared majors in STEM, and who had enrolled in public two-year colleges. In this study, students of color included the following groups: Black, Hispanic/Latino, Multi-Ethnic, and Asian. In this research, these students accounted for 1,372 cases. The next section provides an overview of the variables employed in this study.

Variables

Four outcome variables were explored in this study. Each outcome variable represented a different measure of an academic integration scale. The items were collected on a 3-point scale, including ever, sometimes, and often. They reflected whether students talked with faculty about academic matters outside of the classroom, met with academic advisors about academic plans, worked on coursework at the school library, and used the web to access school library coursework. Hereafter, these variables are referred to as *faculty–student interaction, meeting with advisors about academic plans, studying at the library,* and *Internet for school library usage*. Prior research has shown that these variables illustrated strong construct validity (e.g., Wood, 2012), however, the items illustrated low construct validity with this study's population ($\alpha < .63$). Rather than using the items as a scale, the researchers opted to examine each item as a separate dependent variable.

Five primary predictor variables serve as the core of our analyses. These non-cognitive variables represent standardized scales. They include the following:

- *Math self-efficacy* is a scale item ($\alpha = .91$) that is comprised of students' responses to five questions: "I can do an excellent job on math tests," "I can

understand difficult math texts," "I can understand difficult math classes," "I can do an excellent job on math assignments," and "I can master math class skills." The scale represents student confidence in their academic abilities as it relates to performance in math.
- *English self-efficacy* is a scale item ($\alpha = .92$) that is comprised of students' responses to five questions: "I can understand difficult English texts," "I can understand difficult English classes," "I can do an excellent job on English assignments," "I can do an excellent job on English tests," and "I can master skills in English class." The scale represents student confidence in their academic abilities as it relates to performance in English.
- *Locus of control* is a scale item ($\alpha = .82$) that is comprised of students' responses to four questions: "I can learn something really hard," "I can get no bad grades if I decide to," "I get no problems wrong if I decide to," and "I can learn something well if I want to." The scale ranges on a continuum from an external locus of control to an internal locus. In this study, an internal locus means that the student feels that they have the ability to control their performance in academic contexts.
- *Degree utility* is a scale item ($\alpha = .81$) that is comprised of students' responses to three questions: "I study to get a good grade," "I study to increase job opportunities," and "I study to ensure financial security." This scale represents student perceptions of the worthwhileness (utility) of their academic pursuits as a mechanism to reach their external goals.
- *Action control* is a scale item ($\alpha = .89$) that is comprised of students' responses to five questions: "I remember most important things when I study," "I work as hard as possible when I study," "I keep studying even if material is difficult," "I do my best to learn what I study," and "I put forth best effort when studying." This scale measures the degree of attention students direct towards academic matters.

In addition to the aforementioned variables, all of the models computed employed several control variables. These control variables included: (a) time status, a dichotomous variable indicating whether or not a student was full-time or part-time (part-time enrollment served as the reference category); (b) gender, a dichotomous variable indicating whether or not the student was male or female. Being female served as the reference category; (c) family income, an ordinal variable collected on a 13-point scale, indicating income beginning with: none (coded 1) and $1,000 or less, and ranging to $200,001 or more (coded 13); (d) parent's highest level of education, a variable indicating the highest degree a student's parent had earned. This variable was collected on an 8-point scale, including: less than high school (coded 1); GED or equivalent (coded 2); high school diploma (coded 3); attended or completed a two-year college (coded 4); attended a four-year college without completion: (coded 5); graduated from a four-year college (coded 6); obtained a master's degree (coded 7); and obtained a doctoral or other advanced degree (coded 8); (e) highest level of education (goal), reflecting the highest degree that a respondent wanted to earn. This variable employed the same scale as parent's highest level of education; and (f) high school grade point average, a variable used to account for students' prior academic performance. This variable was collected on a 6-point scale, ranging from 0.00–1.00 (coded 1) to 3.51–4.00 (coded 6). In addition to these variables, this study also controlled for racial/ethnic affiliation. Data employed included Multi-ethnic, Asian/Asian-American, Hispanic/Latino, and Black/African American. The latter served as the reference category.

Analytic Procedure

Prior to advanced analyses, the researchers engaged in exploratory data analysis of the study variables. This included examining means, percentages, standard deviations, and other descriptive data. Further, given the potential interrelationship among the non-cognitive outcomes explored, correlations among the predictor variables were computed. While correlations among predictor variables were strong, ranging from .664 to .744, analyses of variance inflation were well below typical cutoff scores.

Four analyses were conducted to examine the effect of the non-cognitive predictors on the individual measure of academic integration for STEM students. As noted, the dependent variables employed in this study were ordered, reflecting a scale of increasing integration (e.g., never, sometimes, often). Ordinal regression could not be used as the data violated the assumption of parallel lines (O'Connell, 2006; Scott, 1997; Scott & Cheng, 2004). As a result, multinomial logistic regression was employed (see Chen & Hughes, 2004). As such, each outcome category was treated as nominal data. Data were weighted to address complex sampling concerns.

Findings

Faculty–Student Interaction

As noted, the first analysis examined the effect of non-cognitive variables on faculty–student interaction. In the first model, we examined the odds of meeting with faculty never vs. sometimes. Hispanic students had lower odds than Black students, by 46.4%, in meeting with faculty sometimes as opposed to never. A student's goal for their highest level of education was also significantly predictive of faculty interaction differences. Interestingly, students with higher academic goals actually had lower odds (OR = .834, $p <. 05$) of meeting with faculty sometimes as opposed to never. Prior academic achievement, as assessed by high school grades, was positively predictive of meeting with faculty sometimes (OR = 1.696, $p < 001$). In terms of the non-cognitive variables in the model, control expectation was significantly predictive of faculty interaction.

Specifically, the greater students' internal locus, the greater their odds of meeting with faculty sometimes, as opposed to never (OR = 1.665, $p <. 01$). In the second model, we investigated the odds of meeting with faculty often as opposed to never. Multiethnic students had 201% greater odds of meeting with faculty often as opposed to never, in comparison to their Black student peers. Similar to the prior model, Hispanic students had lower odds than Black students of meeting with faculty often as opposed to never, by 68%. An examination of the non-cognitive variables illustrated that students with greater levels of math self-efficacy (e.g., confidence in their math abilities) had significantly lower odds of meeting with faculty often (OR = .604, $p < .01$). Further, locus of control was also found to have a significant effect on whether students met with faculty never vs. often. Students with greater internal locus had greater odds (OR = 2.97, $p < .001$) of meeting with faculty often. The pseudo R^2 for the first analysis was .227.

Meeting with Advisors about Academic Plans

The second analysis examined non-cognitive predictors of meeting with academic advisors. The first model illustrated that men had lower odds than women (by 47.8%) of meeting with advisors sometimes as opposed to never ($p < .05$). This model also showed that English self-efficacy was negatively predictive (OR = .724, $p < .05$) of meeting with advisors about academic plans sometimes. As such, students with lower confidence in their English abilities were more likely to meet with advisors. The second model compared the odds of meeting with an advisor often as opposed to never. As with the first model, men had significantly lower odds than women (by 76.1%) of meeting with advisors often ($p < .001$). This model also showed that family income was inversely predictive of meeting with advisors often (OR = .825, $p < .01$). In terms of non-cognitive variables, this research showed that both math (OR = .676, $p < .05$) and English self-efficacy (OR = .527, $p < .01$) were significant negative predictors of meeting with advisors often. Moreover, locus of control was identified as being significantly predictive of the outcome. This variable was a significant positive predictor of meeting with advisors often (OR = 2.23, $p < .001$). The second analysis accounted for a total pseudo R^2 of .239.

Studying at the Library

The third analysis explored whether non-cognitive variables were predictive of STEM students studying at the library. The first model explored whether students studied at the library sometimes as opposed to never. The model indicated that female students had greater odds, by 96%, of studying at the library sometimes in comparison to their male peers ($p < .05$). This model also showed that family income was inversely related to the outcome (OR = .878, $p < .05$).

In terms of the non-cognitive variables, two illustrated significance. Degree utility was found to be a significant negative predictor of the outcome (OR = .579, $p < .01$). This suggests that STEM students who place greater importance on the utility of their academic endeavors as less likely to use the school library to study. Action control was also identified as a significant variable, illustrating a positive effect on the outcome (OR = 1.410, $p < .05$). As a result, students who direct a higher degree of attention towards academic matters are less likely to use the school library to study.

The second model examined variables that were predictive of studying at the library often as opposed to never. Interestingly, full-time STEM students were 49.6% less likely than part-time STEM students to study in the library often as opposed to never. Possibly, full-timers who are STEM majors have other study locations. The model also indicated that the odds of a male student studying in the library often were 109% greater than that of their female peers. Income was also found to have a significant effect on studying in the library often, however, the relationship was inverse. With respect to non-cognitive outcomes, students with a greater internal locus of control were significantly more likely to study in the library often as opposed to never (OR = 1.708, $p < .01$). Additionally, similar to the first model, students with lower levels of degree utility (belief in the usefulness of their academic endeavors) had lower odds of studying in the library (OR = .472, $p < .001$). This analysis represented a pseudo R^2 of .134.

Using the Internet to Access Library Resources

The fourth (and final) analysis explored variables predictive of STEM students using the Internet to access library resources. The first model indicated that full-time students and men had significantly lower odds of using the Internet to access the school library sometimes as opposed to never, by 57.8% ($p < .05$) and 62.9% ($p < .001$), respectively. This model also indicated that Asian STEM majors had lower odds of using the Internet for school library use than Black STEM majors, by 58.3% ($p < .001$). A similar pattern was seen between Hispanic and Black STEM majors, where Hispanic majors had lower odds (by 57.8%) of using the Internet to access the library ($p < .05$). Family income was found to be a significant but inverse predictor of using the school library sometimes as opposed to never (OR = .833, $p < .01$). Two non-cognitive variables indicated a significant relationship on the outcome. Math self-efficacy was found to be a negative predictor of using the internet to access school library resources sometimes, as opposed to never (OR = .630, $p < .01$). Further, students with greater levels of internal locus had significantly greater odds of using the internet to access school library resources (OR = 2.571, $p < .001$). In the second model, the researchers found that men had lower odds, by 55.7%, of using the internet to access the library often as opposed to never ($p < .01$). Additionally, family income was found to be negatively predictive of the outcome (OR = .762, p

.001). As with the first model, math self-efficacy was found to be negatively predictive of the outcome (OR = .699, $p < .05$). The only other non-cognitive predictor found to have a significant relationship on the outcome was action control. Greater levels of action control were found to result in greater odds of using the internet to access library resources often (OR = 2.460, $p < .001$). This analysis represented a total pseudo R^2 of .251.

Discussion

In general, the models all have predictive utility, accounting for 23% to 25% of the variance in the outcome. The only exceptions were the models for studying at the library, which accounted for only 13.4% of the variance in the outcomes. In three of the four models, math self-efficacy illustrated a negative effect on differences in academic integration measures between respondents indicating never vs. often. English self-efficacy was a nonsignificant predictor in most of the analyses, except with respect to meeting with academic advisors. In that model greater levels of self-efficacy were negatively predictive of differences in academic integration. As a whole, the analyses indicated that self-efficacy more often led to lower levels of academic integration, particularly math self-efficacy. In essence, this means that greater levels of confidence in one's math abilities are more likely to inhibit positive academic patterns (e.g., meeting with faculty, talking with advisors, using the school library). Possibly, students with greater levels of confidence in these areas feel that they have the ability to succeed without support from faculty, advisors, and use of campus resources.

In most models, locus of control was a significant positive predictor of academic integration measures. Locus of control accounted for differences between students indicating never vs. sometimes on faculty–student interaction and using the Internet to access school library resources. This concept also served to demarcate distinctions between students marking never vs. often on all models except for using Internet to access library resources. Given this, results from this study indicate that students with a greater internal locus of control are more likely to have positive academic integration than their counterparts with lower external loci. Degree utility had little effect on any of the models, except for students using the library to study. In all cases, greater levels of utility were negatively predictive of distinctions between students studying at the library never vs. sometimes and sometimes vs. often. Similarly, action control illustrated limited significance in most analyses. Students with greater levels of action control were more likely to study at the library sometimes as opposed to never and to use the Internet for school library usage often vs. never. This relationship is seemingly logical, as greater attention directed towards one's academic endeavors could be manifested by studying and using library resources. With this in mind, the next section presents some recommendations derived from this study.

Recommendations for Research and Practice

It is important for community college faculty and counselors to be more intentional and strategic in their programmatic efforts. Findings from this study suggest that students who report higher levels of self-efficacy are less likely to take advantage of academic opportunities. Many programs have an explicit focus on building self-efficacy in math and science. However, as the data here have illustrated, there may also be a need to reinforce the importance of academic integration experiences. For instance, *all* students need to engage with faculty, meet with advisors, and use campus resources. Communicating the value of these experiences to students may be even more important for those with greater confidence in their academic abilities.

It is especially important for faculty at community colleges to participate in professional development opportunities that train and encourage them to be more strategic in engaging students. Such trainings should focus on strategies building students' sense of internal control and fostering a focus on their academic endeavors. As a result of such training, faculty must communicate to students (through engagement practices) that they are in control of their academic futures. While this is not to say that issues of racism, discrimination, and external pressures (e.g., familial responsibilities, outside encouragement, work/life balance) cannot impede their success. However, findings around locus of control illustrate that a sense of greater personal control can be a strong facilitator of positive integration experiences. One strategy for fostering an internal locus of control may relate to the development of problem solving skills. For example, STEM program counselors and academic advisors can discuss the importance of identifying and evaluating potential courses of action when confronted with issues (i.e., course, major, and faculty selection). In doing so, they can reinforce the notion that students must evaluate their options and make decisions in their best interest. Similarly, faculty can take a similar approach in the classroom by providing students with choices in the selection of assignments, small groups, lab dates, etc. These efforts must reinforce a notion of personal control over academic matters and their career trajectories.

For researchers interested in further exploring this line of inquiry, more scholarship that disaggregates by ethnic group is needed to explore within group differences. A key limitation of this study is that the unweighted sample size was too small for within group analyses. The socialization and cultural experiences of students within the same ethnic group can have grave consequences on their academic success. Moreover, additional research is needed on commuters or non-traditional age students and its impact on their academic integration. These students usually have a more difficult time engaging academically because of off campus living circumstances, employment obligations, limited access and interaction with other students, and the institution's lack of commitment to providing opportunities specific to their needs. Thus, better understanding the nuanced effects of non-cognitive outcomes on integration for these students is needed. Also, further exploration of non-cognitive factors and their impact on academic integration are warranted. Only five measures were used in this study; however, research also shows the importance of other non-cognitive outcomes such as sense of belonging, academic self-concept, and intrinsic interest (to name a few). Future studies should investigate how these constructs add to academic integration experiences for STEM students of color.

The success and increase in participation of students of color in STEM fields has important implications for the nation's ability to maintain a competitive edge within the global market economy. As such, it is becoming increasingly important for institutions of higher education, particularly community colleges, to enhance their STEM production. This is particularly important considering the essential role community colleges play in providing a pathway into four-year colleges and universities. Although the data from this study are admittedly limited to a small sample of students of color in a single cohort, it does provide some compelling information that could potentially be used to further enhance the success of students of color in community colleges.

References

Abd-El-Fattah, S. M. (2005). The effect of prior experience with computers, statistical self-efficacy, and computer anxiety on students' achievement in an introductory statistics course: A partial least squares path analysis. *International Education Journal, 5*(5), 71–79.

Aguayo, D., Herman, K., Ojeda, L., & Flores, L. Y. (2011). Culture predicts Mexican Americans' college self-efficacy and college performance. *Journal of Diversity in Higher Education, 4*(2), 79–89.

Bandura, A. (1977). Self-efficacy: Toward a unifying theory of behavioral change. *Psychology Review, 84*(2), 191–215.

Bandura, A. (1986). *Social foundations of thought and action: A social cognitive theory.* Englewood Cliffs, NJ: Prentice Hall.

Barber, B. (2011). *Characteristics of students placed in college remedial mathematics: Using the ELS 2002/2006 data to understand remedial mathematics placement* (unpublished doctoral dissertation). Arizona State University, Tempe.

Bean, J. P. (2005). Nine themes of college student retention. In A. Seidman (Ed.), *College student retention: Formula for student success* (pp. 215–244). Westport, CT: Praeger.

Bean, J., & Eaton, S. B. (2001). The psychology underlying successful retention practices. *Journal of College Student Retention, 3*(1), 73–89.

Bean, J. P., & Metzner, B. S. (1985). A conceptual model of nontraditional undergraduate student attrition. *Review of Educational Research, 55*(4), 485–540.

Bong, M. (2001). Role of self-efficacy and task-value in predicting college students' course performance and future enrollment intentions. *Contemporary Educational Psychology, 26*(4), 553–570.

Beginning Postsecondary Students Longitudinal Study. (2009a). Second follow-up (BPS:04/09). *STEM majors, base year by race/ethnicity, for first institution sector (level and control) 2003–04 (public 2-year).* Washington, DC: U.S. Department of Education, National Center for Education Statistics.

Beginning Postsecondary Students Longitudinal Study. (2009b). Second follow-up (BPS:04/09). *Students who entered STEM left these fields by spring 2009 by race/ethnicity, for first institution sector (level and control) 2003–04 (public 2-year).* Washington, DC: U.S. Department of Education, National Center for Education Statistics.

Brown, S. D., Tramayne, S., Hoxda, D., Telander, K., Fan, X., & Lent, R. W. (2008). Social cognitive predictors of college students' academic performance and persistence: A meta-analytic path analysis. *Journal of Vocational Behavior, 72,* 298–308.

Chen, C. K., & Hughes, J. (2004, May 26). Using ordinal regression model to analyze student satisfaction questionnaires. *IR Applications,* 1–13. Retrieved from http://www3.airweb.org/images/irapps1.pdf

CCSM. (2012). Overview of the CCSM. *Minority Male Community College Collaborative.* Retrieved from http://interwork.sdsu.edu/sp/m2c3/ccsm/

Chemers, M. M., Hu, L., & Garcia, B. F. (2001). Academic self-efficacy and first-year college student performance and adjustment. *Journal of Educational Psychology, 93*(1), 55–64.

Choi, N. (2005). Self-efficacy and self-concept as predictors of college students' academic performance. *Psychology in the Schools, 42,* 197–205.

DeWitz, S. J., & Walsh, W. B. (2002). Self-efficacy and college student satisfaction. *Journal of Career Assessment, 10,* 315–326.

Essien-Wood, I., & Wood, J. L. (2013). Academic and social integration for students of color in STEM: Examining differences between HBCUs and non-HBCUs. In R. T. Palmer, D. C. Maramba, & M. Gasman (Eds.), *Fostering success of ethnic and racial minorities in STEM: The role of minority serving institutions* (pp. 116–129). New York, NY: Routledge.

Faison, A. C. (1993). *The effect of autonomy and locus-of-control on the academic achievement of Black male community college students* (doctoral dissertation). Available from ProQuest Dissertations and Theses database (UMI No. 9315460).

Flowers, A. M. (2012). Academically gifted Black male undergraduates in engineering: Perceptions of factors contributing to their success in a historically Black college and university. In R. T. Palmer & J. L. Wood (Eds.), *Black men in college: Implications for HBCUs and beyond* (pp. 163–175). New York, NY: Routledge.

Fries-Britt, S., Burt, B., & Franklin (2012). Establishing critical relationships: How Black males persist in physics at HBCUs. In R. T. Palmer & J. L. Wood (Eds.), *Black men in college: Implications for HBCUs and beyond* (pp. 71–88). New York, NY: Routledge.

Gore, P. A. (2006). Academic self-efficacy as a predictor of college outcomes: Two incremental validity studies. *Journal of Career Assessment, 14,* 92–115.

Hagedorn, L. S., & Dubray, D. (2010). Math and science success and nonsuccess: Journeys within the community college. *Journal of Women and Minorities in Science and Engineering, 16*(1), 31–50.

Hardy, D. E., & Katsinas, S. G. (2010). Changing STEM associate's degree production in public associate's colleges from 1985 to 2005: Exploring institutional type, gender, and field of study. *Journal of Women and Minorities in Science and Engineering, 16*(1), 7–30.

Hirschy, A. S., Bremer, C. D., & Castellano, M. (2011). Career and technical education (CTE) student success in community colleges: A conceptual model. *Community College Review.* doi:10.1177/0091552111416349

Ingels, S. J., Pratt, D. J., Rogers, J. E., Siegel, P. H., & Stutts, E. S. (2004). *Educational Longitudinal Study of 2002: Base year data file user's manual* (NCES 2004-405). Washington, DC: U.S. Department of Education, National Center for Education Statistics.

Kane, M. A, Beals, C., Valeau, E. J., & Johnson, M. J. (2004). Fostering success among traditionally underrepresented student groups: Hartnell college's approach implementation of the math, engineering, and science achievement (MESA) program. *Community College Journal of Research and Practice, 28,* 17–26.

Lenaburg, L., Aguirre, O., Goodchild, F., Kuhn, J. (2012). Expanding pathways: A summer bridge program for community college STEM students. *Community College Journal of Research and Practice, 36,* 153–168.

Lester, J. (2010). Women in male-dominated career and technical education programs at community colleges: Barriers to participation and success. *Journal of Women and Minorities in Science and Engineering, 16*(1), 51–66.

Majer, J. M. (2009). Self-efficacy and academic success among ethnically diverse first-generation community college students. *Journal of Diversity of Higher Education, 2*(4), 243–250.

Mason, H. P. (1998). A persistence model for African American male urban community college students. *Community College Journal of Research and Practice, 22*(8), 751–760.

Mesa, V. (2012). Achievement goal orientations of community college mathematics students and the misalignment of instructor perceptions. *Community College Review, 40*(1), 46–74.

Nevarez, C., & Wood, J. L. (2010). *Community college leadership and administration: Theory, practice, and change.* New York, NY: Peter Lang.

Ngo, B. (2006). Learning from the margins: The education of Southeast and South Asian Americans in context. *Race, Ethnicity, and Education, 9*(1), 51–65.

Ngo, B., & Lee, S. J. (2007). Complicating the image of model minority success: A review of southeast Asian American education. *Review of Educational Research, 77*(4), 415–453.

O'Connell, A. A. (2006). *Logistic regression models for ordinal response variables.* Thousand Oaks, CA: Sage.

Pajares, F., & Schunk, H. D. (2001). Self-beliefs and school success: Self-efficacy, self-concept, and school achievement. In R. Riding & S. Rayner (Eds.), *Perception* (pp. 239–266). London, England: Ablex. Retrieved from http://www.des.emory.edu/mfp/PajaresSchunk2001.html

Palmer, R. T., Davis, R. J., & Peters, K. A. (2008). Strategies for increasing African Americans in STEM: A descriptive study of Morgan State University's STEM programs. In N. Gordon (Ed.), *HBCU models of success: Successful models for increasing the pipeline of Black and Hispanic students in STEM areas* (pp. 129–146). New York, NY: Thurgood Marshall College Fund.

Palmer, R. T., Davis, R. J., & Thompson, T. (2010). Theory meets practice: HBCU initiatives that promote academic success among African Americans in STEM. *Journal of College Student Development, 51*(4), 440–443.

Ramos-Sánchez, L., & Nichols, L. (2007). Self-efficacy of first-generation and non-first-generation college students: The relationship with academic performance and college adjustment. *Journal of College Counseling, 10*(1), 6–18.

Scott, L. J. (1997). *Regression models for categorical and limited dependent variables.* Thousand Oaks, CA: Sage.

Scott, L. J., & Cheng, S. (2004). Regression models for categorical outcomes. In M. Hardy & A. Bryman (Eds.), *Handbook of data analysis* (pp. 259–284). Thousand Oaks, CA: Sage.

Solberg, V. S., O'Brien, K., Villarreal, P., Kennel, R., & Davis, B. (1993). Self-efficacy and Hispanic college students: Validation of the College Self-efficacy Inventory. *Hispanic Journal of the Behavioral Sciences, 15,* 80–95.

Solberg, V. S., & Villarreal, P. (1997). Examination of self-efficacy, social support, and stress as predictors of psychological and physical distress among Hispanic college students. *Hispanic Journal of Behavioral Sciences, 19*(2), 182–201.

Starobin, S. S., & Laanan, F. S. (2010). From community college to PhD: Educational pathways in science, technology, engineering, and mathematics. *Journal of Women and Minorities in Science and Engineering, 16*(1), 67–84.

Starobin, S. S., Laanan, F. S., & Burger, C. J. (2010). Special issue on community colleges. *Journal of Women and Minorities in Science and Engineering, 16*(1). doi:10.1615/JWomenMinorScienEng.v16.i1

Thompson, M. D. (2001). Informal student-faculty interaction: Its relationship to educational gains in science and mathematics among community college students. *Community College Review, 29*(1), 35–57.

Tinto, V. (1975). Dropouts from higher education: A theoretical synthesis of recent research. *Review of Educational Research, 45*(1), 89–125.

Tinto, V. (1988). Stages of student departure: Reflections on the longitudinal character of student leaving. *Journal of Higher Education, 59*(4), 438–455.

Tinto, V. (1993). *Leaving college: Rethinking the causes and cures of student attrition* (2nd ed.). Chicago, IL: University of Chicago Press.

Torres, J. B., & Solberg, V. S. (2001). Role of self-efficacy, stress, social integration, and family support in Latino college student persistence and health. *Journal of Vocational Behavior, 59,* 53–63.

United States Government Accountability Office (US GAO). (2005). *Higher education: Federal science, technology, engineering, and mathematics programs and related trends*. Washington, DC: Author.

Vuong, M., Brown-Welty, S., & Tracz, S. (2010). The effects of self-efficacy on academic success of first-generation college sophomore students. *Journal of College Student Development, 51*(1), 50–64.

Wood, J. L. (2011, October 13). Falling through the cracks – An early warning system can help Black males on the community college campus. *Diverse Issues in Higher Education,* 24. Retrieved from http://diverseeducation.com/article/16561

Wood, J. L. (2012). Black males in the community college: Using two national datasets to examine academic and social integration. *Journal of Black Masculinity, 2*(2), 56–88.

Wood, J. L., & Palmer, R. T. (in press). Academic achievement and the community college: Perspectives of Black male students on the importance of 'focus'. *Journal of College Student Affairs*.

Zajacova, A., Lynch, S. M., & Espenshade, T. J. (2005). Self-efficacy, stress, and academic success in college. *Research in Higher Education, 46*(6), 677–706.

Zimmerman, B. J. (2000). Self-efficacy: An essential motive to learn. *Contemporary Educational Psychology, 25*(1), 82–91.

6

STEMMING THE TIDE

Psychological Factors Influencing Racial and Ethnic Minority Students' Success in STEM at Community Colleges

Terrell L. Strayhorn, Michael Steven Williams, Derrick L. Tillman-Kelly, and Marjorie Dorimé-Williams

It has long been recognized that a college degree leads to greater lifetime earnings and gainful employment opportunities (Baum & Payea, 2004). Attending college gives students the skills and knowledge necessary for meaningful participation in democracy and prepares them for today's knowledge- and technology-driven labor force. Yet, recent shifts in the economy have led to an increased need for workers to assume jobs in science, technology, engineering, and mathematics (STEM) fields. President Barack Obama has even implemented an "Educate to Innovate" campaign to improve participation and performance of students in STEM. This campaign includes efforts from the federal government, foundations, non-profits, and science and engineering societies to help students excel in subjects like math and science, in hopes of restoring the United States to its position as a world leader in these areas. Recall that in 2006 the United States ranked 21st (out of 30) in science among developed countries and 25th (out of 30) in mathematics. To achieve our national goals, many more students must *enroll in* and *graduate from* college. College enrollment serves a critical function in preparing America's future workforce, especially in STEM fields.

Although increased college-going rates are important for all racial/ethnic groups if the country is to reach President Obama's 2020 college completion goal, particular attention must be given to students from historically underrepresented ethnic backgrounds, as these students represent an ever-increasing proportion of the nation's population and an "untapped resource," as noted by Kahlenberg (2004). Despite population growth among these groups, racial differences in college student participation continue to persist. Approximately 63% of students who graduate from high school go on to higher education, although rates vary by type of institution (e.g., two- vs. four-year, public or private; National Center for Higher Education Management Systems, 2012). In fact, national reports indicate that upwards of 54% of Native Americans, 51% of Latinos, and 44% of Blacks begin their careers at community colleges (CCs; Knapp, Kelly-Reid, & Ginder, 2011). And while almost half of underrepresented students start at CCs, far too many fail to ever earn a degree (Boggs, 2011), and departure rates are dramatically higher in STEM fields.

Still, community colleges serve several major functions in the tapestry of American higher education. For example, CCs provide much needed access to higher education at a substantially lower cost than traditional four-year institutions. In 2009, 44% of all undergraduate students were enrolled in CCs (American Association of Community Colleges, 2012). Today, there are over 1,700 CCs that offer degrees in fields such as business and accounting, nursing and healthcare, and teacher certification, to name a few (U.S. Department of Education, 2011).

Community colleges also serve a major credentialing function in the United States as they grant certificates, associate's degrees, and even provide vocational job training. Recently, there have been a number of new ventures to increase credentialing and training in STEM fields at CCs. Businesses, four-year colleges, and for-profit organizations all have established meaningful collaborations with CCs to offer STEM focused programs and job training in critical areas. For example, General Motors' partnership with Brookhaven College of the Dallas Community College District provides specialized workforce training to meet the company's unique product knowledge and skill requirements. As another example, The Ohio State University (OSU) and Columbus State Community College are partners in the "Future Scientists of Ohio" initiative that grants full-tuition scholarship to students who plan to earn an Associate of Science in any STEM field, leading to completion of a Bachelor of Science degree in a STEM-related field at OSU.

CCs have also improved STEM participation by encouraging student engagement in educationally purposeful activities and providing vocational counseling and guidance. Previous scholars have noted that students at CCs may not recognize their potential to excel in STEM fields (Strawn & Livelybrooks, 2012); educators, in such instances, play an important role in nurturing their STEM interests, encouraging them to pursue STEM opportunities, and advising them through course selection and transfer options. In this way, CCs have the potential to develop scientific literacy, educate future scientists and engineers, and prepare future technicians (Boggs, 2010).

Existing research tends to focus on (a) history, (b) the transfer function and articulation agreements, (c) and open access, but more work is needed that focuses on racial and ethnic minorities in STEM fields at CCs. The small numbers of individuals being trained to join the STEM workforce continues to be an issue of national concern. The role that CCs play in preparing students to take on these roles is vital for our future economic well-being as well as expanding STEM education for underrepresented minority students.

Purpose

The purpose of this chapter was to examine the influence of background and social psychological factors on racial/ethnic minority students' success in STEM at two-year community colleges. Specifically, we analyzed data from a national sample of students using the Community College Student Experiences Questionnaire (CCSEQ). Our approach sought to identify racial/ethnic differences in CC STEM students' experiences and outcomes, as well as to discern informal and formal experiences, as proxies for social psychological factors, that influence the success of such students at our nation's community colleges.

What We Know from Research

American higher education is characterized by unparalleled diversity that distinguishes it from other systems across the globe. The diversity of American higher education is reflected in the variety of institutions that comprise the system. For instance, there are over 4,300 colleges and universities in the United States; over half of these are two-year CCs. Much of what is known about two-year CCs focuses on their historic purpose, mission, and the establishment of such institutions. For example, CCs are typically open-access (i.e., without strict admissions requirements) and designed to provide low- or no-cost postsecondary education to local citizens or residents and to promote economic development (Laanan, 2001; Thelin, 2004). CCs also tend to cost much less than four-year institutions (Tinto, Russo, & Kadel, 1994). Recent data from the U.S. Department of Education (2011) suggests that two-year public institutions cost $7,703, on average, compared to $15,014 for four-year public universities (see Table 6.1). National statistics also indicate that CCs educate a relatively large number of racial/ethnic minorities (REMs) who aspire to earn associate's degrees or transfer to a four-year institution.

Although relatively large numbers of REMs enroll at CCs each year, some students report significant challenges that, without intervention, can undermine their success and lead to dropout. For instance, despite low tuition, some

TABLE 6.1 Total Tuition, Room and Board for Full-Time Students at Two-Year Public Institutions

Year	Constant $	Current $
2005–2006	7,003	6,492
2006–2007	7,166	6,815
2007–2008	7,073	6,975
2008–2009	7,568	7,568
2009–2010	7,629	7,703

Note: Constant dollars refer to 2008–2009.
Source: U.S. Department of Education, National Center for Education Statistics. (2011). Digest of Education Statistics, 2010 (NCES 2011-015), Chapter 3.

REM families still shoulder enormous financial burdens that can compromise one's commitment to earning a two-year degree or intensify the need to work, which may distract from time spent studying or attending class (Tinto et al., 1994). And although approximately 50% of Black and Latino men who enter postsecondary education begin at two-year CCs (Chenoweth, 1998), less than 20% complete their associate's degree or go on transfer to a four-year institution (Flowers, 2006). In fact, a recent headline in the *Chronicle of Higher Education* read, "Persistence, not access, is problem for 2-year students," based on a report by Policy Analysis for California Education (see Esters & Mosby, 2007). Authors found that approximately 25% of CC students left school after their first semester; success rates were even lower for African American males.

Prior research on REMs at CCs consists of several streams of scholarly inquiry. First, studies have shown that CC students often struggle to balance academic, work, and family responsibilities (Al-Habeeb, 1990; Patterson, 1993; Payne, 2006). For instance, Al-Habeeb (1990) interviewed students, faculty members, and staff to understand why students left a CC in Oregon. Results suggest that part-time, low socioeconomic (SES) students with family and career responsibilities were negatively affected by the programs and services provided at the college.

A second stream of scholarship focuses on the transfer function of two-year CCs. Historically, "2-year colleges have long been touted as agencies for the democratization of opportunity in higher education" (Lucas, 1996, p. 41). Generally, two-year colleges cost less, are less selective in their admissions, and educate large numbers of REMs who might not otherwise access higher education. Traditionally, two-year colleges aim to serve those who are less academically prepared for college (Dougherty, 1992) and those who desire to take developmental courses before transferring to a four-year college.

A small handful of studies provide empirical evidence about REM students' adjustment and satisfaction at two-year CCs (Flowers, 2006; Laanan, 1995; Strayhorn, 2011). For instance, Flowers (2006) analyzed data from a nationally representative sample of African American men in their first year of college and found that Black men at two-year institutions reported higher levels of academic and social integration than their same-race male counterparts at four-year institutions. This is important to note as the weight of empirical evidence suggests that academic and social integration are consistent predictors of satisfaction and retention decisions (Strayhorn, 2012; Tinto, 1993), even at CCs.

In light of recent policy decisions and efforts to broaden minority participation in high-need professions such as STEM fields, CC leaders have established new and revised existing programs to recruit and retain REMs in STEM. For instance, Oakton Community College in Skokie, Illinois offers special STEM classes that provide students with research and design experience, tutoring for supplemental instruction in rigorous STEM courses, a summer bridge program for new students, and student–industry–teacher simulations to nurture one's career aspirations. While promising, there are relatively few studies that test the efficacy of these programs in achieving positive outcomes for REMs at CCs.

A careful review of the literature leads to several major conclusions. For instance, while we know a good deal about the role CCs can play in broadening minority participation in STEM fields, we know considerably less about the success of REMs in STEM at CCs and specific factors that influence their satisfaction in college. This is the gap addressed by the study upon which this chapter is based. Before presenting the results of our analysis, the study's design and sample are described in the next section.

What We Did in the Study

This chapter is based upon a secondary analysis of data from the 2004–2005 administration of the CCSEQ housed in the Center for the Study of Higher Education at the University of Memphis.

Data Source

Since we were interested in the factors that influence REMs' success in STEM at CCs, we needed data from REMs majoring in STEM at one of our nation's more than 2,000 two-year institutions. We settled on using data from the CCSEQ, which consists of 191 items designed to elicit information about the quality and quantity of students' experiences at CCs. Based, in part, on the same philosophy as the College Student Experiences Questionnaire (CSEQ) developed by Pace (1984), the CCSEQ was developed with the belief that the more effort students expend in using institutional resources and opportunities, the more they benefit. And since we were primarily interested in what students do or *could do* to succeed in STEM, the CCSEQ database was deemed appropriate for our work.

Sample

The analytic sample consisted of 2,311 students who had taken at least one course in STEM (e.g., science, math) as measured by the survey. We reduced the sample in this way because the CCSEQ does not include an item about academic major (given that very few two-year colleges require students to declare a major per se). A majority of the sample was White (71%), female (61%), 22 years old or younger (54%), and spoke English as a native language (87%). Table 6.2 presents a full summary of our sample.

Data Analysis

Data analysis proceeded in three steps. First, descriptive statistics were used to calculate means and standard deviations for all dependent and independent

TABLE 6.2 Description of the Analytic Sample ($N = 2,311$)

Characteristics		%
Race		
	White	71
	Hispanic/Latino	12
	Black/African American	8
	Asian/Pacific Islander	6
	Native American	1
Gender		
	Male	39
	Female	61
Age		
	18–19 Years (or younger)	20
	20–22 Years	34
	23–27 Years	18
	28–29 Years	18
	40–55 Years	9
	Over 55 Years	1
English native language		
	Yes	87
	No	13

variables. Second, mean difference tests were used to determine if CCSEQ factors influencing students' success in STEM differed by race. Third, we used analysis of covariance (ANCOVA) to test for racial differences in CC STEM students' experiences and outcomes, controlling for a set of potentially confounding factors. For ease of reading, we footnoted the statistical results in the next section.

What We Learned

Analysis of survey results revealed several significant differences by race among CC students. For instance, Blacks in STEM at CCs were more satisfied with their collegiate experiences than Latinos in STEM.[1] And, consistent with previous research, Whites in STEM at CCs were more satisfied with their collegiate experiences than their Latino counterparts.

Not only do community college STEM students differ by race/ethnicity in terms of their overall satisfaction with college, but we found racial differences among STEM students at CCs in terms of their perceived learning gains—that is, what they felt they learned as a result of college. For instance, we were surprised to learn that Latino students report greater gains than Whites in "understanding of mathematical concepts," and that difference persisted despite controls for potentially confounding variables. In other words, even when White and Latino STEM students at CCs devote the same amount of time to studying, teaching their peers, and seeking help from instructors, Latino STEM students report greater gains in understanding of math than their White counterparts. And since we know from prior research that this is not typically the case when students graduate from high school or enter CCs, our results may point to promising practices (e.g., tutoring, help-seeking, studying) for reducing the racialized "STEM achievement gap" among students. We'll return to this point in the next section.[2]

Given our focus on STEM, we analyzed our survey data to estimate racial differences in CC STEM students' perceived gains in "understanding of the role of science and technology," controlling for a battery of potentially confounding factors. Interestingly, results suggest that CC STEM students do not differ by race in terms of their perceived learning about the role of science and technology (contrary to what we found about understanding of math).[3] This opens up a new set of questions for us to consider in the next section regarding CC STEM students' learning in science and related fields.

So, What Can We Do?

Recall that the purpose of this chapter was to examine the influence of background and social psychological factors on racial/ethnic minority students' success in STEM at two-year community colleges. Using data from the Community College Student Experiences Questionnaire (CCSEQ), we conducted analyses to identify racial/ethnic differences in CC STEM students' experiences and outcomes, as well as to discern informal and formal experiences, as proxies for social psychological factors, that influence the success of such students at our nation's community colleges. Findings from our study suggest several major conclusions that deserve mention.

First, Latino STEM students at CCs report lower satisfaction with their collegiate experiences than Whites and Blacks. Indeed, they reported the lowest satisfaction among all racial and ethnic groups. This is a cause for concern because Latinos are the fastest growing racial/ethnic group in the United States. As such, any future college completion and attainment goals must include intentional and targeted supports for these students to be successful. We encourage CC educators and administrators to take creative and intentional steps to improve the Latino student experience through targeted programming in several ways. For example, there is a theoretical connection between expectations and satisfaction. To address this link, community college agents could administer the College Students Expectations Questionnaire (CSXQ) and compare those results to the CCSEQ to identify areas of congruence and incongruence and direct energies accordingly. Next, CC students tend to be more satisfied when their identity is readily reflected in the campus community (Lane, 2003). This means that efforts to hire and recruit Latino faculty and staff could have an important pay-off by creating a more welcoming, and thus more satisfying campus climate for these students.

Second, although Latino STEM students at CCs report lower satisfaction than Blacks and Whites with their collegiate experiences, they report greater learning gains in terms of math understanding. This is surprising as we know from previous reports (Adelman, 1999) that this is not typically true when students move from high school to college. However, our findings may suggest that when Latino students invest appropriate time and energy toward understanding math, are taught in productive classroom settings, and have access to institutional support services, they may learn just as much, if not more, than their diverse peers. In response to this finding, we encourage CC administrators to consider the impact of institutional support on STEM students' success. Again, CC administrators should also be aware of the diverse needs of students both in and out of the classroom setting, especially when seeking to encourage REM students to persist in STEM fields. Promoting high expectations among faculty, staff, and administrators for REM STEM students can continue to promote high academic achievement for these students.

Lastly, we were surprised to learn that race/ethnicity accounted for so little of the variance in CC STEM students' experiences and perceived learning outcomes. Race accounted for only 12% of the total variance, on average, in students' self-rated learning gains. And although all of us, as critical race theorists, believe strongly that "race matters" (West, 1993) in American society, we were initially confounded by the statistical evidence suggesting that race has a relatively small effect on CC STEM students' experiences and outcomes—much smaller than the effect of, say, transfer status and frequency of teaching one's peers. Let's be clear, we are not saying that our findings support the claim that race does not matter; nor do they suggest that we now live in a post-racial society where one's race has little impact on his/her odds for success in educational contexts. Indeed, nothing could be further from the truth in our collective opinion. Still, findings from our study may lend support to the idea that CC STEM students may be more similar than different across races in their interests, motivations, and experiences. And other factors, such as social class (i.e., SES), family responsibilities, and financial support, may be more likely to distinguish students' experiences in STEM at CCs.[4] CC educators, then, should consider the information included in this chapter when working with students on campus in both academic and social settings. By taking steps to enhance their satisfaction and conditions that promote student learning, they can … no, we can … take giant steps toward broadening minority participation and "STEMming the tide" for STEM students at America's community colleges.

We close with a few recommendations for future research. Our chapter is based upon a statistical analysis of CCSEQ data from a multi-campus sample of CC students. While useful, our analysis was restricted to factors and variables measured by the instrument, which consequently leads to underspecified models as we know that many more factors affect CC student success that are untapped by the CCSEQ. For instance, the survey includes no measures of SES, high school achievement, or reliable items that reflect aspects of the campus' climate toward diversity. Had such measures been included, our results may have been different. Thus, we advise future researchers to develop new or revise existing instruments like CCSEQ to incorporate such items and to place them on appropriate scales (e.g., 5-point agreement) that make sense to students.

In addition to quantitative techniques using ex-post facto survey designs, or newly constructed surveys to enhance our understanding of REMs at CCs, we recommend future qualitative research. In particular, we recommend that researchers employ qualitative techniques (e.g., interviews, focus groups) that seek to the nuances in the lived experiences of REMs in STEM at CCs, their academic intentions beyond the CC, and their overall STEM aspirations. Additionally, findings from such qualitative studies—for example, ethnographic studies of Latino STEM students or case studies of CC STEM students—should be used to refine existing surveys that can be administered to large samples of students and lead to more generalizable results. Indeed, we strongly recommend the integration of quantitative and qualitative techniques in designing sophisticated mixed methods studies of CC STEM students' experiences. Finally, federal funding agencies like the National Science Foundation and private foundations should continue to incentivize such work through competitive grant programs and collaborative partnerships with businesses and institutions.

Notes

1. ANOVA results were statistically significant ($F[5, 2275] = 2.706$, $p < 0.05$), with Blacks in STEM at CCs ($M = 3.92$, $SD = 1.46$) scoring higher than Latinos ($M = 3.58$, $SD = 1.22$) and Whites ($M = 3.84$, $SD = 1.24$) scoring higher than Latinos ($M = 3.55$, $SD = 1.22$) in terms of college satisfaction.
2. ANCOVA results were significant ($F[5, 2081] = 3.93$, $p < 0.001$), with Latinos in STEM at CCs ($M = 2.73$, $SD = 0.96$) reporting greater perceived gains in "understanding of mathematical concepts" than Whites ($M = 2.45$, $SD = 0.97$), controlling for all covarying effects in the analysis. Covariates included sex, transfer status, time spent studying, applying science to other areas, and the frequency of teaching peers or seeking help from instructors.
3. ANCOVA results were statistically *insignificant*, $F(5, 2080) = 0.492$, $p > .05$, controlling for confounding factors. Covariates included sex, transfer status, time spent studying, applying science to other areas, and the frequency of teaching peers or seeking help from instructors.
4. In our analysis, the effect size for race ($n^2 = 0.009$) was much smaller than that for transfer status ($n^2 = 0.027$) and frequency of teaching peers ($n^2 = 0.028$).

References

Adelman, C. (1999). *Answers in the toolbox: Academic intensity, attendance patterns, and bachelor's degree attainment*. Washington, DC: U.S. Department of Education, Office of Educational Research and Improvement.

Al-Habeeb, A. M. (1990). Equal access to the problem of attrition in a community college: A case study. *Dissertation Abstracts International, 52*(01), 62A.

American Association of Community Colleges. (2012). *Fast facts*. Retrieved from http://www.aacc.nche.edu/AboutCC/Pages/fastfacts.aspx

Baum, S., & Payea, K. (2004). *Education pays 2004*. New York, NY: The College Board.

Boggs, G. R. (2010). Growing roles for science education in community colleges. *Science, 329*, 1151–1152.
Boggs, G. R. (2011). Community colleges in the spotlight and under the microscope. *New Directions for Community Colleges, 156*, 3–22.
Chenoweth, K. (1998). The road not taken. *Black Issues in Higher Education, 14*(26), 24–27.
Dougherty, K. (1992). Community college and baccalaureate attainment. *Journal of Higher Education, 63*(2), 188–214.
Esters, L. L., & Mosby, D. C. (2007). Disappearing acts: The vanishing Black male on community college campuses. *Diverse Issues in Higher Education, 24*(14), 45.
Flowers, L. A. (2006). Effects of attending a 2-year institution on African American males' academic and social integration in the first year of college. *Teachers College Record, 108*(2), 267–286.
Kahlenberg, R. D. (2004). *America's untapped resource: Low-income students in higher education.* New York, NY: Century Foundation Press.
Knapp, L. G., Kelly-Reid, J. E., & Ginder, S. A. (2011). *Enrollment in postsecondary institutions, fall 2009; graduation rates, 2003 & 2006 cohorts; and financial statistics, fiscal year 2009* (NCES 2011-230). U.S. Department of Education. Washington, DC: National Center for Education Statistics.
Laanan, F. S. (1995). Making the transition: Understanding the adjustment process of community college transfer students. *Community College Review, 23*(4), 69–84.
Laanan, F. S. (2001). Accountability in community colleges: Looking toward the 21st century. In B. K. Townsend & S. Twombly (Eds.), *Community colleges: Policy in the future context* (pp. 57-76). Westport, CT: Ablex.
Lane, J. E. (2003). Studying community colleges and their students: Context and research issues. In M. C. Brown, II & J. E. Lane (Eds.), *Studying diverse institutions: Contexts, challenges, and considerations* (pp. 51–67). San Francisco, CA: Jossey-Bass.
Lucas, C. J. (1996). *American higher education: A history.* New York, NY: St. Martin's Press.
National Center for Higher Education Management Systems. (2012). College-going rates of high school graduates — directly from high school. Retrieved August 21, 2012, from http://www.higheredinfo.org/dbrowser/index.php?measure=32
Pace, C. R. (1984). *Measuring the quality of college student experiences.* Los Angeles: University of California Center for the Study of Evaluation.
Patterson, E. J. (1993). Factors influencing community college students' transfer to a baccalaureate degree program. *Dissertation Abstracts International, 54*(07), 2437A.
Payne, T. L. (2006). Attrition in an urban southeastern community college: A case study. *Dissertation Abstracts International, 68*(02A), 447.
Strawn, C., & Livelybrooks, D. (2012). A five-year university/community college collaboration to build STEM pipeline capacity. *Journal of College Science Teaching, 41*(6), 47–51.
Strayhorn, T. L. (2011). Traits, commitments, and college satisfaction among Black American community college students. *Community College Journal of Research & Practice, 35*(6), 437–453.
Strayhorn, T. L. (2012). *College students' sense of belonging: A key to educational success.* New York, NY: Routledge.
Thelin, J. R. (2004). *A history of American higher education.* Baltimore, MD: Johns Hopkins University Press.
Tinto, V. (1993). *Leaving college: Rethinking the causes and cures of student attrition* (2nd ed.). Chicago, IL: University of Chicago Press.
Tinto, V., Russo, P., & Kadel, S. (1994). Constructing educational communities: Increasing retention in challenging circumstances. *Community College Journal, 64*(4), 26–29.
U.S. Department of Education, National Center for Education Statistics. (2010). *The condition of education 2010.* Washington, DC: U.S. Government Printing Office.
U.S. Department of Education, National Center for Education Statistics. (2011). *The condition of education 2011* (NCES Report No. 2010-081). Washington, DC: U.S. Government Printing Office.
West, C. (1993). *Race matters.* New York, NY: Vintage Books.

7

THE PROPENSITY TO AVOID DEVELOPMENTAL MATH IN COMMUNITY COLLEGE

A Focus on Minority Students

Bobbie Everett Frye, James E. Bartlett, II, and Kelly D. Smith

Developmental (remedial) education is perhaps the most challenging and important problem facing community colleges (Bailey, 2009). Developmental education is designed to provide students who lack prerequisite skills the means to remediate these skills and to progress to college level coursework in the deficit areas (Bailey, Jeong, & Cho, 2010). Community colleges have "open-door" admittance policies which allow for students that are not prepared for college level work to enroll in college programs. As such, community college students often need remediation and academic assistance to succeed in college (Boylan & Saxon, 1999).

Indeed, remediation is an expensive repair to the educational pipeline and the cost has increased over time. Earlier studies indicated that remediation costs public colleges and universities more than a billion dollars annually (Breneman & Harlow, 1998). However, a more recent study calculated the annual cost of remediation at $1.9–$2.3 billion in community colleges and an additional $500 million at four-year colleges (Strong American Schools, 2008). Clearly, these data indicate that the cost of developmental education programming is concentrated in community colleges.

Achieving the Dream (ATD, 2010), a national community college student success initiative which targets low income and under-represented populations, found that 62% of full-time students in community colleges needed developmental math. Further, 34% need developmental English and 35% need developmental reading (Clery & Topper, 2008). Data analyses of ATD colleges have revealed consistent patterns of entering students who require developmental education, and who experience low completion rates in developmental courses (Achieving the Dream, 2010; Bailey et al., 2010). In comparison to the ATD colleges, a National Educational Longitudinal Study (NELS) examined students from eighth grade in 1988 to the year 2000. The researchers found that 58% of those students who attended a community college took at least one remedial course, 44% took between one and three courses, and 14% took more than three remedial courses (Attewell, Domina, Lavin, & Levey, 2006).

In addition, using a propensity score matching technique that takes into account student differences, Attewell and colleagues (2006) found that attending a two-year college affected one's likelihood of remediation. The probability of remedial placement was 11% higher for students in two-year institutions compared to similar students in four-year institutions. In two-year colleges, taking two or more remedial math courses lowered the likelihood of graduation by 3%. However, remediation in writing and reading improved chances of graduating for two-year college entrants. In sum, there was evidence that successful remediation in two-year colleges improved student outcomes. Most students took 1 or 2 remedial courses in the first year of college. This chapter demonstrates the usefulness of propensity score matching to control for self-selection bias and the practical application of its use in an educational setting.

Why Use Propensity Score Matching?

It is common to analyze the impact of educational interventions using two groups of students, those exposed to the intervention and those not exposed. Yet results are limited in that the students are not typically randomly selected into experimental and control groups. Non-random selection implies that the two groups of students may be very different on key factors that affect the results of analyses through self-selection bias and other differences. Propensity score matching (PSM) is a technique designed to simulate an experimental design, controlling for selection bias and creating almost equivalent experimental and control groups on key indicators. Comparisons of student outcomes using propensity matching, has been used to yield less biased results than are derived using simple comparisons (Rosenbaum & Rubin, 1984; Titus, 2007).

Thus study will use a program developed by Strong American Schools (SAS; Parsons, 2000) that matches treated students to the first closest match in the control group, based on propensity scores. Propensity score matching is used to derive treatment and non-treatment groups matched on the propensity or conditional probability to avoid developmental math courses. The advantage of using propensity score modeling is that the method addresses selection bias common in program evaluation impact studies when the treatment and non-treatment groups are not randomly selected. The propensity to take remedial courses, the dependent variable, was determined by developmental math enrollment behavior in the first year of college.

The Purpose

The purpose of this study is to determine if there is a difference in community college outcomes (grade point average, credits earned, credentials earned, transfer, persistence) between two groups: (a) an avoider (control) group of minority students who avoid developmental math coursework in the first year of enrollment and (b) a non-avoider (treatment) group of minority students who attempt developmental math coursework in the first year of enrollment. Bearing the aforementioned in mind, two primary research questions guided this study:

1. Is there a difference in student and behavioral characteristics between minority developmental students who within one year of enrollment avoid developmental math coursework compared to students who do enroll in one or more developmental math courses?
2. Is there a difference in college level outcomes between minority developmental math students who within one year of enrollment avoid developmental coursework compared to students who enroll in one or more developmental math courses?

Avoidance and Completing the Developmental Education Sequence

Exploratory research has expanded the conceptual understanding of developmental course completion rates to developmental sequence completion rates. The *course sequence* refers to the series of courses in the subject area a student needs to complete in order to fulfill the developmental sequence. In a study of over 256,000 students in 57 ATD colleges, Bailey and colleagues (2010) examined referral, enrollment, and completion in developmental math and reading. They found, between 33% and 46% of students actually completed the required sequence of courses. However, as expected, developmental completion rates were negatively related to remediation levels in math and reading. In addition, about a third of students avoided developmental coursework by never enrolling in any developmental courses. Interestingly, the results also indicated more students exited their developmental course or sequence because they did not enroll in the first or subsequent course than because they failed or withdrew from a course in which they were enrolled (Bailey et al., 2010). Overall, men, older students, part-time students, and students in vocational programs were less likely to progress through developmental coursework.

In addition, Bailey and colleagues (2010) found 21% of students placed in developmental math and 33% of students placed in developmental reading skipped the developmental sequence in reading and math, enrolled in, and successfully completed the gatekeeper course, 12% and 32%, respectively. One measure of program effectiveness is the extent to which passing the developmental course or sequence leads to successful completion of the gatekeeper or gateway course in the subject area. Since some students skipped the recommended remediation and successfully completed the gatekeeper course, the researchers concluded skipping developmental coursework in some instances was a wise strategy for developmental students.

Bostian (2008) hypothesized that some students transfer from a community college to a four-year university ostensibly to avoid remedial coursework, which in practice carries no college credit and can delay completion (Bettinger & Long, 2005; Martorell & McFarlin, 2007). The purpose of his research study was to determine the extent of "remedial avoidance" for two first-time cohorts of 910 community college students. He selected two groups of students for analysis, the avoider group and the non-avoider group. He matched them based on the conditional probability they were members in one group versus another, i.e., propensity score matching. Another objective of the study was to determine, after matching, were there differences in the two groups of student's outcomes who transferred to a local four-year university (Bostian, 2008).

Bostian's (2008) study of remedial avoidance found that 14% ($n = 127$) of the total 910 students in the study population avoided remediation and transferred to the four-year university. However, 178 of the original cohort did not avoid remediation, transferred to the four-year institution and composed the non-avoider group. The student background and academic characteristics differed for the two groups among several independent variables such as placement test scores, age, and grade point average. However, once the differences in students were controlled using propensity score matching, there were no detectable differences among the 89 matched students in the non-avoider and avoidance groups. The groups fared about the same in terms of degree completion and grade point averages. The researcher concluded that students do avoid remediation, but those that have the tenacity and intention to transfer to the university achieved comparable results to the non-avoider group. While the sample size was small (Bartlett, Kotrlik, & Higgins, 2001) and may be inadequate, the study filled an important gap in the literature by following students after transfer to one institution. However, we know little about the extent to which outcomes vary in significant ways among different student groups based on student background, and academic characteristics, goals and intentions. The next section examines student outcomes of minority students in a grant funded program designed to increase representation in STEM pathways.

Student Progress and the Minority Student in Community Colleges

The Louis Stokes Alliances for Minority Participation (LSAMP) Program was established in 1991 by the National Science Foundation (NSF) to develop strategies to increase the quality and quantity of minority students who successfully complete baccalaureate degrees in science, technology, engineering, and mathematics (STEM), and who continue on to graduate studies in these fields (Clewell, 2006). The NSF grant selection process is competitive but two- and four-year institutions submit proposals and outline strategies that will increase the progress and successful completion of underrepresented minority populations in STEM areas. The Urban Institute was charged with evaluating the national program and surveyed LSAMP participants who graduated from college between 1992 and 1997. The data collected provided student demographics and outcomes of interest that aligned with the goals and objectives of the LSAMP program such as undergraduate grade point averages and enrollment in and the successful completion of those programs (Clewell, 2006).

As previously discussed, a LSAMP survey was administered to measure and document the impact of LSAMP among participants. Of salience to this chapter, student outcomes were compared for two groups of students those who began their studies at a community college and those that did not. Specifically, almost 10% of students began their studies at a community college. However, community college starters were demographically different than those students who did not attend community colleges in three areas. They were significantly more likely to be male (63% vs. 49%), Hispanic (54% vs. 36%), and to have a mother whose education level was less than high school completion (25% vs. 15%). Despite demographic differences, program outcomes were comparable for the two groups of students in terms of undergraduate grade point averages, enrollment in STEM graduate programs and the successful completion of STEM graduate degrees (Clewell, 2006).

As a result, the evaluators recommended that researchers focus on the role of community colleges in preparing students for STEM careers paying close attention to the linkages among community colleges and four-year institutions. The evaluators also determined more research is needed to inform the field about student success outcomes and the progress of students interested in STEM degrees. In particular, longitudinal cohort studies are needed that track progress and successful completion of minority students in STEM areas (Clewell, 2006).

The Problem

Limited research has focused on the role of community colleges in the production of racial and ethnic minority students in the STEM pipeline. This is particularly troublesome, given that minority students comprise 45% of the community college population (AACC, 2008). Given the high incidence of developmental math needs among community college students, developmental math avoidance, course-taking patterns, and subsequent outcomes are the focus of this chapter on minority students. Leinbach and Jenkins (2008) suggest subpopulation analyses in an institutional context will further the understanding of factors affecting the success of student groups. Bearing this in mind, this chapter will contribute to the literature in three areas: (a) by examining racial and ethnic developmental math students to determine differences in student demographics and academic backgrounds as a function of developmental math course-taking behavior; (b) measuring the difference in program outcomes in an institutional context by focusing on the role of the community colleges in the facilitation of racial and ethnic minorities in STEM areas; and (c) demonstrating the utility and feasibility of propensity score matching as an evaluative technique in educational settings.

Assessment, Placement, and Lack of Consistency

Typically, developmental students are determined by a placement test score and the cutoff point for developmental versus college level placement is determined by departments that varies from institution to institution or by statewide policies (Bailey et al., 2010; Oudenhoven, 2002; Perin, 2005). As a result, there is little or no consensus on what constitutes "college ready" among different community colleges. The number of developmental courses required in each subject area is sometimes referred to as the remediation level. For example, Remediation Level 1 in Math would indicate a student enrolls in and passes one developmental course (the highest level developmental Math) in Math before proceeding to college-level math. However, some colleges vary in terms of developmental levels and offer one level of developmental math while some offer as many as four levels of developmental math (Bailey et al., 2010). In addition, some programs of study exempt students from higher level courses in the recommended remediation levels while vocational certificates may completely exempt students from developmental math. Moreover, there is variation among developmental programs in terms of voluntary versus mandatory attendance. While Alfred and Lum (1988) found a negative effect between remediation and subsequent course grades for students in voluntary remedial programs, they, as did Boylan, Bliss, and Bonham (1997), found a positive effect for persistence and graduation rates for students in mandatory remedial programs.

The lack of consistency in developmental policies and practices led Bailey and colleagues (2010) to refer to developmental requirements as "mazes" through which students navigate. In practice, the accumulation of developmental credits does not accrue to college credits earned and can delay college completion (Martorell & McFarlin, 2007). Hence, the value of developmental courses is uncertain for students and developmental courses are "systematically avoided" among some students (Grubb & Cox, 2005). Developmental course requirements can become frustrating for students who must meet them but who may be surprised to learn they are not earning college credits towards their degree or are not earning credits that are recognized at most four-year institutions.

Student Outcomes in Developmental Education

Researchers investigating the effectiveness of developmental education have used a variety of outcomes as indicators that the developmental student has remediated successfully (Bettinger & Long, 2005; Calcagno & Long, 2008). In general, outcomes of interest vary from short-term to long-term outcomes. Developmental education programs are designed to provide students who lack prerequisite academic skills the means to remediate skills and progress to college level coursework in the deficit subject areas such as reading, English and math (Bailey et al., 2010). Therefore, one measure of program effectiveness is the extent to which students remediate skills, progress to college level coursework and pass the gatekeeper course or courses in the areas of deficiency. Sustained progress, persistence, and completion of short-term outcomes in theory should lead to the completion of long-term outcomes. Due to the nature of the college experience, longitudinal studies are analyzed to answer the long-term questions regarding the effects of remediation.

Bettinger and Long (2005) conducted a statewide data analysis of community college students in Ohio. They examined over 13,000 first-time traditional aged, 18- to 20-year-old students for a 5-year period from 1998 to 2003. Full-time and part-time remedial students were compared to full-time and part-time non-remedial students in terms of credits completed, transfers and credentials earned. The study found that full-time remedial students earned 5.4 fewer credits, were 15 percent more likely to have left college without a two-year degree and 3.6% more likely to have left college without a four-year degree, compared to full-time non-remedial students. In addition, part-time remedial students were less likely to complete two or four-year credentials or transfer compared to part-time non-remedial students. However, part-time remedial students completed more credit hours than part-time non-remedial students.

Bettinger and Long (2005) concluded the following:

> Although a simple comparison suggests that remedial placement has a negative impact on students, it masks the fact that students are not randomly placed in remediation. Better-prepared students are less likely to be placed in remediation and they also do better in college. Thus, simply comparing remedial students with non-remedial students is an unsatisfactory way to establish the true effects of remediation.
>
> (p. 23)

Since the comparison of remedial students to non-remedial students technically does not provide information on the effectiveness of remedial education, the researchers used differences among placement policies in Ohio and distance from high school to college to compare outcomes of students placed into developmental education to similar students not placed into developmental education due to placement policies. The methodology is known as a regression discontinuity design and is used to control for self-selection bias. Similar students were compared to each other and differences among students were controlled in the study. The researchers found positive outcomes for students placed in math remediation. The students placed into math remediation were 15% more likely to transfer by spring 2003 than similar students not placed into remediation. However, the study found no positive effects for students placed in English remediation (Bettinger & Long, 2005).

In Florida, researchers Calcagno and Long (2008) examined community college students who had taken a college placement test. They compared students placing just below the cutoff placement score to those right above the cutoff score in another utilization of regression discontinuity designs. In order to establish consistency in the study, Florida community colleges were excluded that did not use the placement test scores to determine remediation. Out of 130,000 community college students, around 75% or 98,146 students were tracked for a total of 6 years. The three cohorts consisted of first-time, associate's degree-seeking students from 1997–2000. All three cohorts were tracked for 6 years. Short-term outcomes examined included the completion of the subsequent college level course in the remedial subject area and fall-to-fall persistence. Long-term outcomes examined included the completion of a credential, certificate and or associate's degrees, transferring to a four-year university and total credits earned in remedial and college level courses.

They found students just below the cutoff score were more likely to persist from fall to fall than students right above the cutoff score, remedial and non-remedial students, respectively. Also, developmental math students just below the cutoff score earned more total credits over 6 years than students right above the cutoff score. A closer examination of total credits revealed the differences in total credits among remedial and non-remedial students held for remedial credits but did not hold for college level credits or credits toward a degree. In other words, there was no difference between the two groups in college credits earned. In addition, remediation in math had no effect on the completion of college level math or the completion of subsequent English or math courses. Moreover, among the remedial group, researchers found a negative effect in associate's degrees earned and transfer out rates. Adult students enrolled in remedial courses experienced slight positive effects.

In summary, the studies that evaluate the effectiveness of developmental education yield mixed and inconclusive results. Constraints are placed on populations of students examined (Bettinger & Long, 2005). Regression-discontinuity designs are used to control for self-selection bias. First-time students taking a placement test and placing right above or right above placement levels is a population of interest utilizing the regression discontinuity designed studies (Bettinger & Long, 2005; Calcagno & Long, 2008). Differences among placement policies in Ohio were used by Bettinger and Long (2005) to compare outcomes of students placed into developmental education to similar students not placed into developmental education due to placement policies. In general, a few good studies provide information about students who are on the cusp of remediation or of 18- to 20-year-old students. Since most researchers have focused on students right above and below placement, there is a limited amount of information for students who do not place on the cusp of remediation but are in need of many levels of remediation in more than one subject area (Bailey, 2009). In his examination of the studies of developmental education effectiveness including Calcagno and Long (2008) and Bettinger and Long (2005), Bailey (2009) concluded not all students in remediation benefit from or need developmental coursework but more research is needed. The developmental education literature is growing in terms of statewide analyses. However, longitudinal research on developmental students is needed at the institutional levels to inform practice and policy (Leinbach & Jenkins, 2008).

In developmental education, given the extent of costs in time and resources taxed on students, faculty, and the institutions, community colleges need to evaluate developmental education using the best approach feasible and practical. While research has expanded at the national level, this quantitative study examines students in developmental education at one institution and seeks to create a matched group of minority students for evaluative purposes.

Theoretical Framework

Drawing on persistence theory, the theory that underpinned this study was the model of non-traditional undergraduate student attrition developed by Bean and Metzner in 1985. Bean and Metzner's (1985) theory has been used to study factors that impact the decisions to stay or leave postsecondary education among non-traditional students. This theory indicates that non-traditional undergraduates have at least one of the following factors: A student is considered non-traditional if the student is more than 24 years of age, enrolled part-time, and/or a commuter student. The conceptual model also includes: (a) background variables (i.e., age, enrollment status, residence, personal goals, educational goals, high school performance, ethnicity, gender); (b) academic variables (i.e., study habits, advising, absenteeism, major certainty, course availability); (c) environmental variables (i.e., finances, hours of employment, outside encouragement, family responsibilities, opportunity to transfer); (d) academic outcomes (i.e., grade point average); (e) psychological outcomes (i.e., utility, satisfaction, goal commitment, perceived stress); (f) intent to leave; (g) dropout; and (h) social integration.

Bean and Metzner (1985) posit that there is a relationship between academic and environmental factors. When both factors are favorable, the students are likely to stay in college. On the other hand, if both factors of academic and environmental factors are poor, students leave college. In addition, they argue positive environmental factors outweigh poor academic outcomes and students persist even when facing academic hurdles if they are externally supported. External support such as financial assistance can provide needed resources to help the student persist. Similarly, Bean and Metzner (1985) argue there is a relationship between grade point average and psychological factors for non-traditional students. If both grade point average and psychological factors are favorable, the student persists. On the other hand, if both grade point average and psychological factors are poor, students will leave college. In addition, they argue poor psychological factors outweigh a poor grade point average and students will more likely persist than not persist if satisfied with their educational experience. Satisfaction with the educational experience includes satisfied with policies and practice in place that support the student's academic experiences.

Conceptual Framework and Independent Variables

The conceptual framework (Bean & Metzner, 1985) establishes a relationship between academic and environmental factors while controlling for background factors. Favorable academic and environmental factors lead to positive psychological outcomes. Positive psychological outcomes lead to short-term and long-term outcomes. Background factors are important in the attrition model and are measured using demographics such as race, age, and gender. Race categories are a set of dummy variables coded as White, Hispanic, American Indian, Native Hawaiian, mixed race, Asian, Black, and International Student. Age is a continuous variable and gender is dichotomous. Educational goals are defined using two categories of majors: a college transfer degree and vocational associate's degree (Bahr, 2009). There are three types of enrollment status: enrollment status in the first term, first time in college and late entry. Enrolled in 12 or more credit hours is considered full-time enrollment. First time in college is dichotomous and defined as a first time college student. Late entry is dichotomous and defined as 25 years or older (Bean & Metzner, 1985).

There are four categories of math proficiency:

MAT 050—whole numbers, fractions, decimals (4 levels below college);
MAT 060—applications of fractions, decimals, percents, ratio and proportion, order of operations, geometry (3 levels below college);
MAT 070—signed numbers, exponents, simplifying expressions, solving equations, graphing, factoring (2 levels below college);
MAT 080—factoring, rational equations, systems of equations, graphing quadratic and radical functions, complex numbers, variation (1 level below college).
Accuplacer Reading and Sentence Skills scores are continuous and measure proficiency in reading and English.

There are two categories of reading proficiency and four categories of English proficiency:

ENG 080—Focuses on standard conventions of written English. Goals include effective sentence structure, paragraph creation, etc. (2 levels below college);

ENG 085—Develops proficiency with reading and writing at the college level. Critical reading skills are developed: a combination of ENG 080/RED 080 (2 levels below college);

ENG 090—Focuses on the development of effective paragraphs and adhering to standard conventions of English (1 level below college);

ENG 095—Develops comprehension and analysis skills in relation to evaluating college texts: a combination of ENG 090 and RED 090 (1 level below college);

RED 080—Emphasis is placed on vocabulary, comprehension, and reading strategies Teaches students to identify main ideas and supporting details (2 levels below college);

RED 090—Improves reading and critical thinking skills. Topics include vocabulary enhancement; extracting implied meaning; analyzing purpose, tone, and style; and drawing conclusions and responding to written material (1 level below college).

Academic outcomes, the next factors in the model, represent course enrollments which reflect commitment and course attempts that are conducive to persistence for developmental students. Four dichotomous variables represent course academic commitment or integration, enrollment in the student success course, in developmental reading, in developmental English, and in college level English. The course commitment variables are dichotomous variables for each of the subject areas. Academic integration is also defined as academic grade point average and retention. Academic grade point average is defined as the first term grade point average and is a continuous variable. Retention is defined as returned second term and the variable is dichotomous. The one environmental pull factor (Bean & Metzner, 1985) a Pell recipient in first term, a dichotomous variable that represents financial support in college.

There are two categories of outcomes in the model, short-term and long-term outcomes. The short-term outcomes reflect credit attempts, completions, and successful grades of C or better in developmental math (MAT 080, MAT 070, MAT 060, MAT 050), developmental reading, developmental English, and college level English. In this model the short-term outcomes lead to the long-term outcomes. The long-term outcomes are credential attainment or transfer to another institution and final grade point average.

Methodology and Analytical Framework

Study Population

The population consists of the fall 2006 new student cohort at one urban community college. Student progress and outcomes are tracked for 5 years. In order to provide institutional context, the Carnegie Classification is a very large, two-year associate's public institution, with higher part-time enrollment and urban serving multi-campuses located in the Eastern region of the United States. The new student cohort consists of 3,625 students out of approximately 20,000 students. 1,321 students comprised associate's degree seekers referred to developmental math and about half 54% ($N = 642$) were minority students. For purposes of this study, a minority student is defined as Black, Asian American, Hispanic, American Indian, and International Student.

The student record dataset includes demographics, course completion data, and transfer information available through the National Student Clearinghouse. Through participation in the Completion by Design Initiative, a project of the Tides Center, funded by the Bill and Melinda Gates Foundation, the college was provided with a raw data extract. The extract includes individual college submitted data provided to the Completion by Design partners and a series of derived measures provided by the Community College Research Center (CCRC), a data partner in the Completion by Design Initiative. In addition, the raw data extract was merged with internal college data sets in order to create additional measures for this study. Because Integrated Postsecondary Education Data Systems (IPEDS) cohorts represent a select group of degree seeking students (Leinbach & Jenkins, 2008), the study does not restrict the analysis to IPEDS (first-time, full time, certificate or degree declared) classifications but examines all full- and part-time students in associate's degree programs. All of the students completed college placement tests and were referred to developmental math coursework prior to beginning their studies.

In the study population, students are assessed in three different areas—English, Reading, and Math—and are often placed in more than one area of developmental course requirements. Students choose to take the developmental coursework; no mandatory policy is in place but students are blocked from classes in the developmental need areas until meeting the developmental requirement. General Education courses also have requirements that will block students from enrolling in them. As a result, mandatory placement is somewhat enforced through course prerequisites specifically in the general education courses.

Selection of Pre-Treatment Covariates and Pre-Screening

The researchers began by identifying covariates variables at pre-treatment that can bias comparison studies (Dehejia & Wahba, 2002; Rosenbaum & Rubin, 1984; Titus, 2007). Covariate variables are the independent variables with the highest degree of influence on the dependent variable. In general, appropriate independent variables explain differences in group membership that are constant over time (Rosenbaum & Rubin, 1983, 1984). In an educational environment, constants include student demographics and academic variables (Jenkins,

Zeidenberg, & Kienzl, 2009; Rojewski, Lee, & Gemici, 2010). These variables were used in the logistic regression to generate a propensity score (ranging on a scale from 0.0 and 1.0).

Missing data analysis indicated that Native Hawaiian and Mixed Race categories were not populated and the two groups were removed from the analyses. An analysis indicated that 18 records were missing on the reading and sentence test scores values so list-wise deletion was used (see Hair, Black, Babin, & Anderson, 2010). Before proceeding to analysis, the data were examined for multi-collinearity and outliers. The decision was made to retain all the data.

Descriptive Results

An examination of descriptive results indicated that 642 minority students were in the research study (see Appendix). Of those 642 students, 56.1% were female and 43.9% were male. In terms of age, 66.9% of students were 19 years and under, 18.2% were 20-26, and 14.80% were 27 and older. Students in general had transitioned into college right out of high school, but 14.8% were adult students 27 years old and up. Blacks were 69.3% of the group, and the next largest group (15.6%) was comprised of non-resident alien or international students. Of the remainder, 8.9% were Hispanic, 1.7% American Indian, and 4.5% were Asian Americans.

The majority of students 77.1% ($N = 495$) were first time college students. Pell recipients were almost evenly split between the two groups, and slightly more 52.3% ($N = 336$) were enrolled full time in the first term. The students took placement tests in math, reading and English. Among them, 23.8% placed into the highest level of developmental math, MAT 080, 7.4% placed into MAT 070, almost 17.1% placed into MAT 060, and over half 51.6% placed in to MAT 050 the lowest level of developmental math. Sixty-five percent ($N = 415$) enrolled in at least one developmental math course in the first academic year. Among the students, 63.9% ($N = 398$) also placed into developmental reading and 57.6% ($N = 354$) placed into developmental English.

Analyses

In Table 7.1 (see Appendix) stepwise logistic regression was conducted to determine which independent variables race, gender, age, Pell recipient, enrollment statuses, program declarations, course attempts, retention, grade point average after first term and reading, English and math proficiencies were associated with the dependent variable-avoiding or attempting developmental math in the first year of enrollment. Logistic regression results indicated that the overall model of six predictors (Pell recipient, enrolled full time, attempted a student success course, attempted college English, returned the second term, placed into MAT 050 and Accuplacer sentence skills scores) were statistically reliable in distinguishing between attempting and avoiding developmental math among developmental math students (-2 LL^1 = 814.727), chi-squared = 216.34, $p < .001$, $R^2 = .402$.

The model correctly classified 83% of the cases and explained 40% of the variance in the dependent variable. Wald statistics indicated that 6 variables significantly predict group membership for developmental math attempters. Students who received Pell awards in the first term were 64% more likely to take their developmental math courses than students not receiving Pell awards. Students enrolled in college full time were 150% more likely to take developmental math courses in the first year than were part-time students. Students enrolled in a student success course were 114% more likely to fall in the non-avoider group, and students enrolled in college English courses were 88% more likely to do so. Scores in Accuplacer sentence skills proficiency were the weakest predictor of non-avoidance, .01%. The strongest predictor of developmental math non-avoidance was whether the student returned for a second term and these students were 227% more likely to fall in the non-avoider group. However, students in the lowest level of math proficiency, MAT 050, were 73% less likely to take developmental math than were students in the highest math proficiency level.

T-tests were conducted to determine which independent variables students differed on for the two groups of avoiders or non-avoiders of developmental math courses in the first year. Several categories of independent variables were grouped in terms of similarity. First, demographic variables were examined which included gender, race, enrollment statuses, and receipt of Pell awards. Before matching, avoiders were more likely to be male, $t = 2.55$, $p < .01$ and Black, $t = 3.00$, $p < .01$. Non-avoiders were more likely to be Pell recipients in the first term, $t = -4.03$, $p < .01$ and enrolled full time, $t = -8.69$, $p < .01$. Second, the groups of placement variables were examined for math, reading and English subject areas. The results indicated non-avoiders had higher average scores on reading placement tests, $t = -3.21$, $p < .01$ and sentence skills placement tests, $t = -6.49$, $p < .01$. Non-avoiders were also were more likely to place in MAT 060, $t = -3.60$, $p < .01$ and MAT 070, $t = -3.47$, $p < .01$. In contrast, avoiders were more likely to place in MAT 050, $t = 8.72$, $p < .01$, the lowest level of developmental math courses. Last, the two groups were compared on academic progress variables which included being retained to spring, attempting a student success course, attempting a developmental reading course, attempting a developmental English course, attempting a college level English course and first term grade point average. Non-avoiders were more likely than were avoiders to be retained to spring, $t = -9.73$, attempt a student success course, $t = -4.68$, $p < .01$, attempt a developmental reading course, $t = -3.29$, $p < .01$, attempt a college English course, $t = 9.17$, $p < .01$ and have a higher grade point average, $t = -6.12$, $p < .01$. Effect sizes indicate that the significant variables with the greatest influence in the dependent variables reflect those retained in the initial logistic regression analysis. After propensity matching, one independent variable, attempted developmental reading, was statistically different between the two groups. The effect size was low (.20) and overall indicated a successful match in creating two equivalent groups. To further substantiate a successful match a subsequent logistic regression was conducted and no variables entered the model indicating the match was successful (Rojewski et al., 2010).

After matching, the average means differ between the two groups and the results were positive and significant for developmental math attempters in terms of developmental course outcomes. Overall, non-avoiders attempted and completed higher averages of developmental credits than did the avoiders. The non-avoiders attempted an average of 15 credits compared to an average of 8 credits for the avoiders with a large effect size of 0.64. The differences in credits attempted yielded a striking difference between the two groups after propensity matching. The difference between the two groups is prevalent among all the developmental math courses in attempts, completions, grades of C or better and among college level math with effect sizes ranging from .365 to .755. However, there were little or no significant differences between the two groups in the other developmental subject areas, reading and English, or in the college level counterpart, college level English. College level credits attempted, completed, and completed with C or better did not differ significantly between the two groups and neither did mean grade point average. Completion and transfer outcomes also did not differ between the two groups.

The frequencies of highest educational outcomes after 5 years for the matched groups, $N = 141$. Similar to the t-tests there are few significant differences between the two groups after matching. Of practical significance, the non-avoiders were slightly more likely to transfer to a four-year institution with an award, 2.84% ($n = 4$) compared to 0.71% ($n = 1$), and to transfer to a four-year institution and earn a bachelor's degree 2.13% ($n = 3$) compared to 0% ($n = 0$). In addition, the highest educational outcome for avoiders was either not enrolled 70.92% ($n = 100$) or none 5.67% ($n = 8$) compared to non-avoiders who were not enrolled 63.83% ($n = 90$) or none 6.38% ($n = 9$).

Discussion

To summarize, two research questions were addressed in the study. The first question addressed whether there were differences in student characteristics and course-taking attempts between avoiders and non-avoiders of developmental math in the first year. Before propensity matching, there were significant differences between developmental students who within the first year of enrollment avoided developmental math coursework compared to students who did not avoid developmental math courses. T-tests were conducted to determine which independent variables were significantly different between the two groups of the dependent variable defined as placing into developmental math and attempting or avoiding developmental math courses.

Prior to the propensity match, the groups were statistically different on 14 of the 22 independent variables. After propensity matching, one independent variable, enrolled in developmental reading, was statistically different between the two groups. The effect size was low (.20) and overall indicated a successful match in creating two equivalent groups. To further substantiate a successful match, a subsequent logistic regression was conducted and no variables entered the model indicating the match was successful (Rojewski et al., 2010).

Furthermore, prior to matching, stepwise logistic regression results indicated that the overall model of six predictors (Pell recipient, enrolled full time, enrolled in student success course, enrolled in college English, returned second term, placed into MAT 050 and Accuplacer Sentence Skills Scores) were statistically reliable in distinguishing between attempting and avoiding developmental math among developmental math students (-2 Log Likelihood = 814.727), chi-squared = 216.34, $p < .001$, Nakelkerke R Squared = .402. The key take away is that the logistic regression was useful in explaining the predictors of membership in the two groups. In addition, the analysis demonstrated that the two groups were different on several indicators.

The second research question concerned the differences in outcomes after matching. Propensity matching reduced the bias between the two groups and t-tests were conducted yielding unbiased indicators. Four categories of outcomes were examined: completion of developmental courses, completion of college level courses, grade point average, credential completion outcomes and/or transfer to a four-year university.

The non-avoider group attempted and completed, on average, more developmental credits and more developmental math credits than the avoider group. However, the two groups did not differ in terms of developmental English and reading credits. In addition, among college level math credits the two groups differed significantly. The results suggested completion of developmental math credits led to successful completion of college level math. Grade point average was not significantly different between the two groups after matching. Moreover, minimal and non-significant in both groups, were the completion of credentials and/or transferring to a four-year university.

Conclusion

Developmental education is perhaps the most important issue in community colleges, especially for developmental math. The developmental educational literature provides understanding of developmental students' outcomes (Bailey, 2009; Bailey et al., 2010; Bettinger & Long, 2005; Calcagno & Long, 2008). However, limited research has focused on the role of community colleges in the production of racial and ethnic minority students in the STEM pipeline, although minority students comprise 45% of the community college population (AACC, 2008). To the researchers' knowledge propensity score matching has not been used to study the effectiveness of developmental math minority students in the community college. Leinbach and Jenkins (2008) suggested sub-population analyses in an institutional context will further the understanding of student groups. This chapter contributed to the literature in three areas: (a) by examining racial and ethnic developmental math students to determine differences in college outcomes as a function of developmental math avoidance behavior; (b) by measuring the differences in outcomes in an institutional context and focusing on the role of the community colleges in the facilitation of racial and ethnic minorities in STEM areas and; (c) by demonstrating the use of propensity score modeling to create equivalent avoider and non-avoider groups in terms of explanatory variables of avoidance behavior. A minority student cohort was tracked for 5 years.

Propensity score modeling yielded interesting differences between the two groups (such as enrollment status, Pell awards recipients, attempting a student success course, and persistence) that need further examination. Since the objective of propensity score matching was to create equivalent groups, there was little exploration of some of the initial differences among the students and the extent to which those differences warrant investigation. Initial stepwise logistic regression results, indicated that the overall model of six predictors (Pell recipient, enrolled full time, enrolled in student success course, enrolled in college English, returned second term, placed into MAT 050 and CPT Sentence Skills Scores) was statistically reliable in distinguishing between attempting and avoiding developmental math among developmental math students.

Researchers should explore the extent to which the receipt of Pell awards is related to non-avoidance of developmental math. Bean and Metzner (1985) posit that environmental pull factors are important to retaining students in college. An analysis of outcomes provided little, if any, information about the correlation between financial aid and successful completion of math courses. Also, full time enrollment was another strong indication that students would begin their developmental math coursework in the first year. Moreover, placement into the lowest level of developmental math (MAT 050) precipitated developmental math avoidance and is significant given that these students were in need of the most remediation.

Future research areas that emerged from the study indicated the need to continue to examine the effectiveness of developmental education in the community college. The results revealed motivated and persistent minority students enrolled in developmental math, English, and reading subject areas. Since the study is quantitative, there is no indication of the support or academic systems students needed in their studies that would have facilitated progress through the pathways (Clewell, 2006). In addition, quantitative data reflect what the outcomes were between the two groups but do not explain why students avoided developmental math in the first year.

Developmental math and college level math are the pathways for students pursuing STEM degrees. Interestingly the non-avoidance of developmental math in the first year was a good indicator of developmental math credit enrollment and subsequent enrollment and success in college level math. Less conclusive is whether or not the enrollment in and completion of developmental and college level math credits leads to the completion of substantive awards or credentials.

Implications for Practice

While the developmental education literature is growing and seeks to examine the effectiveness of developmental education, an equally important question to address should regard educational research methodology as a topic of significance. Indeed, Levin and Calcagno (2008) critiqued evaluation studies in the community college and argued that ideal studies utilize an experimental design. The second choice among these researchers would be quasi-experimental designs.

The researchers argued that community college research offices are understaffed and evaluation studies are not appropriate in providing effective information. Institutional researchers need to provide information that institutions can use to improve services to developmental students but spend most of their time generating reports for state and federal agencies. The size of the institutional research office plays a role in the type of research conducted. In most cases, staff are assigned numerous roles in an institutional research office, and experimental designs are not practical (Levin & Calcagno, 2008). However, in developmental education, given the extent of costs in time and resources taxed on students, faculty, and the institutions, community colleges need to evaluate developmental education using the best approach feasible and practical. While research has expanded at the national level, quantitative studies need to examine students in developmental education at the institutional levels.

Propensity matching is a technique designed to simulate an experimental design, controlling for selection bias and creating almost equivalent experimental and control groups on key indicators. Comparisons of student outcomes using propensity matching, has been used to yield less biased results than are derived using simple comparisons (Rojewski et al., 2010). It is critical that researchers apply methodologies that control for selection bias. Resources are scarce and monies need to be allocated to educational programs that are making a difference in student's success and in student's long-term outcomes.

This study applied propensity score matching in order to contribute to evaluative studies. Propensity matching was used in order to illuminate the methodology's use in the evaluative literature and to connect the study to one of the purposes for evaluation in educational settings: to strive for continuous improvement of educational programs. As Grubb (2001) pointed out in his review of the developmental education literature, "colleges must evaluate remedial education using sophisticated techniques such as treatment and control groups; program evaluation and improvement is central to improving remedial student's outcomes" (p 10). If developmental education is not evaluated effectively and extensively, then researchers, practitioners, and policy makers might make decisions on incomplete and biased evidence. It is imperative that researchers evaluate developmental programs and effectively contribute to the success of all community college students.

Note

1. -2LL refers to the -2 Log Likelihood

Appendix A

TABLE 7.1 Demographic Information for 642 New Minority Students in Fall 2006

Category	N	Percent
Gender		
Male	282	43.93%
Female	360	56.07%
Age Range		
19 and Under	430	66.98%
20–26	117	18.22%
27 plus	95	14.80%
Race		
Black	445	69.31%
Asian	29	4.52%
American Indian	11	1.71%
Hispanic	57	8.88%
Non-Resident Alien	100	15.58%
Enrollment Status		
First Time in College	495	77.10%
Not First Time in College	147	22.90%
Full-time in First Term	336	52.34%
Part-time in First Term	306	47.66%
Pell Grant Status		
Pell Recipient in First Term	320	49.84%
Not Pell Recipient in First Term	322	50.16%

Category	N	Percent
Developmental Math Status		
Developmental Math in First Year	415	64.64%
No Developmental Math in First Year	227	35.36%
Developmental Placement		
Math Placement		
MAT 080 (1 level below college)	153	23.83%
MAT 070 (2 levels below college)	48	7.48%
MAT 060 (3 levels below college)	110	17.13%
MAT 050 (4 levels below college)	331	51.56%
Reading Placement		
College Level	198	30.84%
RED 090/ENG 095 (1 level below college)	259	40.34%
RED 080/RED 085 (2 levels below college)	139	21.65%
Missing	7	1.09%
English Placement		
College Level	272	42.37%
ENG 095/ENG 090 (1 level below college)	196	30.53%
ENG 080/ENG 085 (2 levels below college)	168	26.17%

References

Acheiving the Dream. (2010). Process for institutional change in Achieving the Dream. Retrieved from http://www.achievingthedream.org

Alfred, R., & Lum, G. (1988). Remedial program policies, student demographic characteristics, and academic achievement in community colleges. *Community College Journal of Research and Practice, 12*(2), 107–120.

Attewell, P. A., Domina, T., Lavin, D. E., & Levey, T. (2006). New evidence on college remediation. *The Journal of Higher Education, 77*(5), 886–924.

Bahr, P. R. (2009). Revisiting the efficacy of postsecondary remediation: The moderating effects of depth/breadth of deficiency. *Review of Higher Education, 33*(2), 177-205.

Bailey, T. R. (2009). Challenge and opportunity: Rethinking the role and function of developmental education in community college. *New Directions for Community Colleges, 145,* 11–30.

Bailey, T. R., Jeong, D. W., & Cho, S. W. (2010). Referral, enrollment, and completion in developmental education sequences in community colleges. *Economics of Education Review, 29*(2), 255–270.

Bartlett, J., Kotrlik, J., & Higgins, C. (2001). Organizational research: Determining appropriate sample size in survey research. *Information Technology and Learning Journal, 19,* 43–50.

Bean, J. P., & Metzner, B. S. (1985). A conceptual model of nontraditional undergraduate student attrition. *Review of Educational Research, 55*(4), 485–540.

Bettinger, E. P., & Long, B. T. (2005). *Addressing the needs of under-prepared students in higher education: Does college remediation work?* Cambridge, MA: National Bureau of Economic Research.

Bettinger, E. P., & Long, B. T. (2005). Remediation at the community college: Student participation and outcomes. *New Directions for Community Colleges, 129,* 17–26.

Bostian, B. (2008). *Avoiding remedial education: Academic effects on college transfer students.* Retrieved from www.cpcc.edu

Boylan, H., Bliss, L., & Bonham, B. (1997). Program components and their relationship to student success. *Journal of Developmental Education, 20*(3), 2–8.

Boylan, H. R., & Saxon, D. P. (1999). *What works in remediation: Lessons from 30 years of research.* Report prepared for The League for Innovation in the Community College. Retrieved April, 7, 2005, from http://www.beaufortccc.edu/progrm/developmental/assets/documents/whatworks.pdf

Breneman, D., & Harlow, W. (1998). Remedial education: Costs and consequences. Remediation in Higher Education Symposium. *Thomas B. Fordham Foundation Report, 2*(9), 1–57.

Calcagno, J. C., & Long, B. T. (2008, July). *The impact of postsecondary remediation using a regression discontinuity approach: Addressing endogenous sorting and noncompliance.* National Bureau of Economic Research Working Paper Series. NBER Working Paper No. 14194. Cambridge, MA: National Bureau of Economic Research. Retrieved from http://www.nber.org/papers/w14194

Clery, S., & Topper, A. (2008). Keeping informed about achieving the dream data. *Data Notes, 3*(4) July/August.

Clewell, B. C. (2006). *Revitalizing the nation's talent pool in STEM.* Retrieved from www.aacc.nche.edu

Dehejia, R. H., & Wahba, S. (2002). Propensity score-matching methods for nonexperimental causal studies. *Review of Economics and statistics, 84*(1), 151–161.

Grubb, W. N., & Cox, R. D. (2005). Pedagogical alignment and curricular consistency: The challenges for developmental education. *New Directions for Community Colleges, 129*, 93–103.

Grubb, W. N. (2001). *From black box to Pandora's box: Evaluating remedial/developmental education.* Retrieved from http://www.cfder.org/uploads/3/0/4/9/3049955/from_black_box_to_pandoras_box_evaluating_remedial_ordevelopmental_education.pdf

Hair, J. F., Black, W. C., Babin, B. J., & Anderson, R. E. (2010). *Multivariate data analysis* (7th ed.). Englewood Cliffs, NJ: Pearson Prentice Hall.

Jenkins, D., Zeidenberg, M., & Kienzl, G. (2009). Educational outcomes of I-BEST, Washington state community and technical college system, an integrated basic education and skills training program: Findings from a multivariate analysis. (Unpublished report). New York, NY: Community College Research Center.

Leinbach, D. T., & Jenkins, D. (2008). *Using longitudinal data to increase community college student success: A guide to measuring milestone and momentum point attainment.* CCRC Research Tools No. 2. New York, NY: Community College Research Center.

Levin, H. M., & Calcagno, J. C. (2008). Remediation in the community college: An evaluator's perspective. *Community College Review, 35*, 181–207.

Martorell, P., & McFarlin Jr, I. (2007). Help or hindrance? The effects of college remediation on academic and labor market outcomes. *The Review of Economics and Statistics, 93*(2), 436–454.

Oudenhoven, B. (2002). Remediation at the community college: Pressing issues, uncertain solutions. *New Directions for Community Colleges, 117*, 35–44.

Parsons, L. S., (2000). Using SAS software to perform a case control match on propensity score in an observational study. *Proceedings of the Twenty-Fifth Annual SAS Users Group International Conference* (pp. 1166–1171). Cary, NC: SAS Institute Inc.

Perin, D. (2005). Institutional decision making for increasing academic preparedness in community colleges. *New Directions for Community Colleges, 129*, 27–38.

Rojewski, J. W., Lee, I. H., & Gemici, S. (2010). Using propensity score matching to determine the efficacy of secondary career academies in raising educational aspirations. *Career and Technical Education Research, 35*(1), 3–27.

Rosenbaum, P. R., & Rubin, D. B. (1983). The central role of the propensity score in observational studies for causal effects. *Biometrika, 70*(1), 41–55.

Rosenbaum, P. R., & Rubin, D. B. (1984). Reducing bias in observational studies using subclassification on the propensity score. *Journal of the American Statistical Association, 79*(387), 516–524.

Strong American Schools. (2008). Diploma to Nowhere. Retrieved from http://www.deltacost-project.org/resources/pdf/DiplomaToNowhere.pdf

Titus, M. A. (2007). Detecting selection bias, using propensity score matching, and estimating treatment effects: an application to the private returns to a master's degree. *Research in Higher Education, 48*(4), 487–521.

8

MOVING BEYOND THE BARRIER OF MATHEMATICS AND ENGAGING CULTURALLY RELEVANT PEDAGOGY IN THE CLASSROOM FOR RACIAL AND ETHNIC MINORITY STEM STUDENTS IN COMMUNITY COLLEGES

Denise Yull

I still remember my first day as a mathematics instructor at the local community college. I was full of nervous energy and excitement. I was not a novice in the classroom; I did have some teaching experience as an adjunct lecturer of mathematics at a couple of colleges in Buffalo, New York. My entry in the classroom, however, was different this time. I had decided to abandon my lucrative engineering career to earn less than half the salary teaching mathematics. I made my decision with a most altruistic mindset. It was best for my family, and I believed I could make a difference. When I decided to leave private industry for a career in education, I believed that in my position as a mathematics instructor in the community college I might be able to inspire and make a difference in the lives of the students. At the very least, my experience as a Black, female, mechanical engineer might be useful in inspiring racial and ethnic minority students and in particular Black students to consider careers in science, technology, engineering, and mathematics (STEM).

I was motivated by the historic positioning of mathematics as the gateway or barrier for many students who want to major in fields leading to careers in STEM. Racial and ethnic minorities—especially Black students—are singularly marked for failure in community college mathematics courses, as they are barred from success by secondary mathematics course experiences which have inadequately prepared them for college (Mullin, 2011). When I left my career as a mechanical engineer and decided to become a mathematics teacher, I felt that I would be able to provide academic support and mentorship to those academically underserved students in my courses as a strategy to remove mathematics as a barrier to their success. After teaching for a couple of years, I recognized that students in my courses were not passing through the mathematics barriers and that they were failing at the same rates as students of other math instructors. Despite my best intentions, my students were not finding success in the mathematics classes I taught. Once I was introduced to Ladson-Billings (1995, 1997) and the culturally relevant pedagogical practices used in secondary education, my class instruction changed. Instead of being a transmitter of mathematical knowledge, I created an environment in my class where we were a community of cooperative learners of mathematics. I evolved to become an effective mathematics instructor, allowing my students to pass through the gate of mathematics into potential careers in STEM. In this chapter, I share my experiences as I moved from my STEM career to a mathematics instructor with the ultimate goal of informing readers about the effective pedagogical strategies that I found to be useful to improve the learning experiences of racial and ethnic minority community college STEM students.

STEM and Community Colleges

Education theorists are providing special attention to postsecondary education in community colleges, with particular attention toward programs leading students to degrees and/or certificates in areas involving STEM. This focus on STEM is being prompted by the global economic crisis, and there is a concerted effort by policy makers, researchers, and educators to address the shortages in the number of students vying for degrees in STEM (Lowell & Salzman, 2007; Maltese & Tai, 2011; Palmer, Maramba, & Dancy, 2011). The shortage of STEM graduates has been connected to the specter of the United States losing its stature and advantage in the global marketplace, specifically in the area of technological innovation (Lowell & Salzman, 2007; Öztürk, 2007). According to President Obama (2006), "Eight of the nine fastest-growing occupations for this decade will require scientific or technological skills and some form of higher education to fill the jobs of the future" (p. 194). Researchers concur with the president, suggesting that employment opportunities are on the rise for those who have postsecondary education in the STEM fields. Based on the work of Carnevale, Smith, and Strohl (2010), 8 million job openings are projected in STEM-related careers over the next 10 years. The current emphasis on STEM has been fueled by prevailing concerns that in the future there will not be a sufficient number of U.S. citizens prepared to fill STEM-related jobs (Lowell & Salzman, 2007). Reports, such as *Rising Above the Gathering Storm: Energizing and Employing America for a Brighter Economic Future* (2007) warn us of the dire consequences to the United States if we do not significantly increase the number of individuals qualified to work in the STEM areas.

President Obama (2009) has proposed that community colleges be used as the primary site in the effort to increase the number of qualified STEM workers. President Obama set a national goal for the United States to produce five million additional college graduates over the next decade as part of the Educate to Innovate Initiative (2010). The initiative proposes uniquely positioned community colleges as the new postsecondary educational school of choice. As such, community colleges will be where a new generation will begin the quest for STEM-related jobs (Lucore, 2012). In the State of The Union speech, President Obama (2010) called this period of time this generation's "Sputnik moment" in reference to when the Soviet Union launched its satellite and triggered the space race in 1957. Though the Soviet Union was first to place a satellite into space, the United States rallied with investments in research, technology, and education which eventually resulted in putting the first man on the moon. In his declaration, the president calls on Americans to out-educate and out-innovate the rest of the world in the new century (Obama, 2010).

Meeting the projected demand for more STEM skilled workers will be no easy feat, but it will be practically impossible if the educators and job creators within United States do not provide consideration to this country's changing demographics. Racial and ethnic minorities who have generally been left out of the technological boom will make up 45% of the working age population by 2030 (Kirsch, Braun, & Yamamoto, 2007). The U.S. Census projects that 50% of the U.S. population will be non-White (e.g., Asian, Black, Latino/a, or Native American) by 2050. These statistics strongly suggest that racial and ethnic minorities will have to be included if the United States is to meet the increasing workforce requirements for STEM-related jobs. Increasing the number of racial and ethnic minorities in STEM fields starts with making sure they do not get "filtered out" of the process early because they fail to get past the barrier of mathematics.

Mathematics: The Gatekeeper to STEM

Mathematics has a history of being regarded as a gatekeeper discipline for those seeking degrees in STEM. In 1989, the National Research Council reported mathematics as being the gatekeeper/barrier that would let people into or keep them out of STEM careers. Specifically, Pendergast (1989) reports:

> More than any other subject, mathematics filters students out of programs leading to scientific and professional careers. From high school through graduate school, on average we lose half the students due to mathematics each year.
>
> (p. 4)

Years after Pendergast's findings, researchers continue to document the gatekeeping nature of mathematics courses (Crisp, Nora, & Taggart, 2009; Gasiewski, Eagan, Garcia, Hurtado & Chang, 2011), suggesting that mathematics continues to be a major barrier for students seeking to enter careers in STEM and it poses a particularly difficult obstacle for racial and ethnic minority students. Racial and ethnic minority students that enter community colleges with dreams of obtaining degrees in STEM-related fields find that, in many cases, these dreams are deferred and even derailed when they take their first mathematics course (Hagedorn & Purnamasari, 2012; Ryan & Siebens, 2012). In fact, mathematics has been known to serve as an impediment for the dreams and aspirations of many seeking degrees in general (Mullin, 2011). Having worked as a mechanical engineer for nearly 15 years, I knew achieving a STEM degree and getting past the barrier of mathematics was possible. I had succeeded in obtaining a STEM degree and career and I believed it was possible for other racial and ethnic minority students to achieve similar goals. The reality for most racial and ethnic minorities is that they are not gaining access to the STEM fields in the numbers that will translate into President Obama's goal set forth in the Educate to Innovate Initiative (Obama, 2010). The lack of racial and ethnic minority students persisting and graduating with STEM degrees has seemingly always been a problem. However, what has changed is that national interest has spurred the need for more workers that are prepared to fill STEM related jobs. This national directive intersects with the changing demographics of the population to create an urgent need for racial and ethnic minorities to attend college and obtain degrees in STEM.

My Life in STEM as a Mechanical Engineer

I have had an opportunity of spending my entire adult life in STEM careers: first, as a mechanical engineer, and more recently, as a mathematics instructor at a community college. As an engineer, I experienced first-hand how an engineering degree could open possibilities for a satisfying and lucrative career. My first job after completing my engineering degree was with Bell Aerospace Textron. I graduated from college with a Bachelor of Science degree in mechanical engineering and went into a career where I instantly made more money that my father had made in his 30-year career in the military. I had an opportunity to work in research that was linked to the Strategic Defense Initiative (SDI), which was proposed by President Ronald Reagan. The Reagan administration funded research that looked at using ground and space-based systems to protect the United States from attack by strategic nuclear ballistic missiles. This was the so-called Star Wars initiative. While it was ultimately deemed to be a tremendous waste of money (Abrahamson & Cooper, 1993), the program did fund a number of research efforts in the companies who acted as defense contractors, keeping a host of engineers like myself employed. My work at Bell Aerospace ended with the downsizing of the Star Wars program.

My next job was as a design and manufacturing engineer at the Harrison Radiator, an automotive component parts manufacturer, which was a division of General Motors. Over the course of my career at Harrison, I worked in the design and manufacturing of automotive radiators. I experienced my job as an engineer as both interesting and challenging; yet, there was always something missing. The focus of my work at Harrison Radiator was building parts that would keep people cool in the summer and warm in the winter as they drove their automobiles. Though important work, my work did not change the lives of many of the Black children I saw on the east side of Buffalo, New York. While I worked as an engineer, I spent a number of hours working in after-school programs trying to inspire Black children and other racial and ethnic minority students to consider careers in STEM. My motivation came each day when I went to work with 100 or so engineers, yet there were only a paucity of engineers who were racial and ethnic minorities, with Blacks making up the smallest of this group. I wanted the children I worked with to aspire to complete STEM degrees which could lead them to future employment and economic empowerment.

The push to increase the number of racial and ethnic minority students in STEM is not a new quest nor is the call for education reform which would result in a technologically literate workforce. The National Council of Teachers of Mathematics reported in 1989 that reform in mathematical education would be needed to prepare the future workforce of the United States for a technology-based economy. Moses and Cobb (2001) declared mathematical literacy should be considered a civil rights issue, suggesting that racial and ethnic minorities would be subjugated to positions which deny them economic and educational access to the new high-tech economy which would ultimately deny full citizenship.

In the late 1970s, when I graduated from high school, I was in the initial group of students asked to participate in Buffalo Engineering Awareness for Minorities (BEAM), which was being sponsored by General Motors. I was encouraged and mentored to choose engineering as my major in college. My work with racial and ethnic minority students in after school programs and in classrooms had instilled in me a desire to help them access and complete STEM degrees. My interest intersected with a life changing opportunity and I found myself changing careers and becoming a mathematics instructor at a community college in Binghamton, New York.

My New Career: Mathematics Instructor

I walked into my first mathematics class with the full knowledge that I would be teaching a course that could potentially hinder a student from completing their college degree. I also knew that many people in our society suffer from math anxiety or math phobia. According to Ashcraft and Kirk (2001), math anxiety has been conceptualized as "a feeling of tension, apprehension, or fear that interferes with math performance" (p. 1). Prior to becoming a mathematics instructor, I engaged in conversations with a number of people who shared with me their fear and/or dislike of mathematics. Ladson-Billings (1997) suggests that this attitude about mathematics has been cultivated in the United States because, as a culture, we have learned to accept mathematics illiteracy as the norm. Based on the work of Ladson-Billings (1997), we accept the notion of a lack of competencies or abilities in mathematics whereas a confession of one's inability to read would be considered embarrassing and unacceptable in the United States. Ladson-Billings (1997) contends that

> Ours is a nation where no one would readily admit to being unable to read but many proclaim with pride their inability to balance their check book or compute the amount of interest on a loan. Not knowing how to read carries a stigma ... those who cannot read or write will mask this fact using a number of strategies, i.e. pretending they cannot see without their eyeglasses.
>
> (p. 698)

In contrast, when people struggle with mathematics it is attributed to lack of innate ability. To profess that you "can't do math" does not cast the same negative shadow on a person that saying "I can't read" would. Ladson-Billings (1997) suggests that we accept mathematical illiteracy because, "The U.S. cultural beliefs support the idea that either one 'has it' or does not when it comes to mathematical ability, or the way to 'get it' is through genetic inheritance" (p. 699). While inability to read is not accepted as a social norm, mathematical illiteracy is accepted, with many people proclaiming their inability to "do math."

I began my career as a mathematics instructor with the full understanding of the social contract we have made as a culture regarding mathematics. I understood that it was culturally excusable to perform poorly in mathematics classes because of the belief that some students just did not have the ability to "get it." I knew that I would encounter many students who would find mathematics so challenging that they would enter class expecting to fail. A number of students in my first class shared with me that they were nervous, and even afraid of mathematics because of their prior classroom experiences. Prepared with this knowledge, I entered my new profession as a mathematics teacher like a warrior on a quest to provide assistance to students in finding success in math in order to help them overcome their mathematics phobia. I set about teaching mathematics with one of my primary goals being to change these students' perceptions of math and their ability to do math.

Learning to Teach

I began my first college algebra class at the community college by asking the students to write their answers to two questions on index cards. The questions were: (a) What do you like about mathematics? and (b) What worries or concerns do you have about our mathematics class? Both questions elicited the same response, which was generally some version of "I hate math" or "I can't do math." I have started each semester by asking my students these same questions and after nearly 15 years of teaching mathematics, I have found that the negative responses come from students who have had some negative experiences in mathematics in prior classes. These negative experiences had one of two effects on my students, either convincing them they were incapable of learning mathematics or that there was no real need for them to learn mathematics. It was my mission to not only help my students overcome their mathematics phobia, but also to show them that they were more than capable of "doing mathematics." As I began my first semester and entered my first class, I was confident and convinced that a person like myself who had a degree in engineering and many years of experience in that career would be more than prepared to teach a mathematics course at a community college. The introductory college algebra class I taught, according to our mathematics department, was equivalent to the mathematics course most local high schools taught freshman year. My course was one in a series of "developmental mathematics" courses designed to prepare students for college level mathematics. Across the nation, 40% to 50% of students entering community college require mathematics remediation (Byrd & MacDonald, 2005; Hoyt, 1999; McCabe, 2003). With great enthusiasm, I began my course with lectures which mimicked the teaching pedagogies that I had experienced in mathematics classes as a student.

Using Familiar Pedagogy: Drill and Practice

My teaching methods in my early days as a mathematics instructor involved a traditional approach which is often found in many mathematics classrooms (Ladson-Billings, 1995, 1997; Tate, 1995). My lectures typically emphasized drill and practice, repetition, convergent thinking, right-answer thinking, and predictability (Ladson-Billings, 1997). Mathematics, as I presented it, was a series of static disconnected steps to be memorized. I had a very linear approach to teaching mathematics: "Do this first and then you get … and the next step is…." Essentially, I taught mathematics the way I was taught. I had learned mathematics in classrooms where it was expected that students would memorize some fundamental facts (e.g., multiplication tables) and learn to carry out certain procedures that would ultimately result in the correct answer to a math problem. As a student, I was taught using a direct instruction approach where my role was as the recipient of mathematical knowledge given to me by teachers. In my mathematics classes, the lectures promoted memorization, suggesting that this was how you learned mathematics; this was how I learned mathematics. The focus was on learning the content. I just assumed that my one-way, passive approach would lead to my students' success in the math. I was wrong.

My first class was a disaster; I had a 70% fail rate after my first exam. My interaction with my colleagues confirmed that we were all using similar techniques in our mathematics classes. I experienced colleague after colleague suggest that some students will get it and others will not, as they had succumbed to the belief in innate mathematical ability.

But There Is No "Math Gene"

In defense of my colleagues, this belief has deep roots within U.S. society. Uttal (1996) suggested while there is no evidence to support the claim of innate mathematical ability, but many Americans do believe that one can be genetically predisposed to having mathematical aptitude which ultimately hinders the mathematical performance of students. So, if mathematics instructors believe that "the cream will rise to the top" (Martin, 2012; Herrnstein & Murray, 1994; Thernstrom & Thernstrom, 1997, 2004) in a mathematics class, they do not find it disturbing when students fail; even if this fail rate exceeds 70%. I must admit there were times when I acquiesced to this thinking, but these were short lived because I understood that there was not a mathematics gene and my students who were failing were not suffering from a genetic predisposition to fail math. I understood this on a very personal level, because when I was an undergraduate in a mechanical engineering program I had been told by a very prominent professor that I was not engineering material. He suggested that I change my major because he did not believe that I had the aptitude for engineering. My successful graduation and career proved that he was wrong; it was a statement I never forgot.

Rethinking My Pedagogical Approach

After a couple years of dismal results in the mathematics courses I was teaching, I became determined to understand what was occurring in my courses and how I might change the outcome for my students. I needed to change the direction of my courses to truly engage my students. I had to first come to terms with my own teaching. How did my teaching pedagogy influence my student's chances of successfully completing their mathematics courses? By examining and reflecting on my teaching pedagogy and class structure, I recognized that I taught mathematics as if the mathematics classroom was a meritocracy. My experience as a student and my training as an engineer had positioned me to believe that success in a mathematics course was purely related to a work ethic. Based on this experience, I was one of those educators who believed that students would be able to grasp the mathematical concepts I was teaching if they worked hard enough. My approach to teaching was a traditional approach, where I stood at the front of the classroom and lectured, showing the students a number of problems while they sat passively in the classroom taking notes. There was very little interaction from the students even with my occasional pause to ask if they had a question. Clearly, I needed to change my instructional methodologies.

Changing Direction: From Drill and Practice to Cultural Competence

Changing how mathematics is taught has been part of the K-12 reform efforts for a number of years. The National Council of Teachers of Mathematics (NCTM) and the National Assessment of Educational Progress (NAEP) have been reporting on the "gatekeeper" status of mathematics since 1989. Discussions with my colleagues in the mathematics department suggested that most understood that students were being kept from opportunities because of their inability to successfully complete their mathematics classes. Some even used the term *gatekeeper*, but the general consensus was that students failed as a result of the student's lack of effort. I cannot remember ever having a discussion about the impact of a teacher's pedagogical approach to the learning outcomes of a student; my colleagues were consistent with their traditional instructor/content centered approach to instruction. This is consistent with what Mesa, Celis, and Lande (2011) found in their study of teaching mathematics in a community college that there was "a consistent portrait of instruction characterized by an emphasis on traditional instructor/ content centered approach" (p. 37). Even as I taught mathematics using algorithms and symbols that were meaningless to my students, I felt that there had to be a better way. Significantly, the pass rate of students in several mathematics courses was often a point of discussion, and faculty often concluded that their pass rates were dismal. As a result, faculty often contended that the high failure rate was due to students not working hard enough. I recognized that faculty had not discussed ways to change their instructional practices. Based on this formative experiences, I became committed to a deeper understanding of mathematics and I decided to go back to school for a Masters of Arts degree in mathematics.

Ethnomathematics: The Beginning of a Shift in Pedagogy

When I began a master's program in mathematics, I suddenly had a moment of clarity. It occurred to me that when I looked at posters of prominent mathematicians that were displayed proudly on a wall in the mathematics lounge, the images stared back at me, Hilbert, Klein, Gödel, and others, were all White men. The poster evoked a visceral response. I thought about my own experiences as a mathematics student and teacher. I thought about the truths that had become embedded in my psyche that was considered one of the powerful truths of the mathematics profession; that is, mathematics has its roots in Europe. Mathematics had been invented by White European men. Powell and Frankenstein (1997) described this notion as the Eurocentric myth of mathematical knowledge creation. This conceptualization of mathematical knowledge creation is important because with it comes the idea that some are inclined to understand mathematics while others are not (Herrnstein & Murray, 1994; Martin, 2012; Thernstrom & Thernstrom, 1997, 2004). The concept of the "math gene" suggests that only the gifted are predisposed to perform mathematics. In the world of mathematics, most mathematics teachers—particularly at the college level—are White. According to the American Institute of Mathematics, approximately 1,200 doctorate degrees are awarded in mathematics with approximately 8% (96) being from under-represented racial and ethnic minority groups. The unspoken suggestion for those who subscribe to the theory of a math gene is that White people are more predisposed to mathematical talent than racial and ethnic minorities.

The belief that mathematics is acultural or innate is countered by ethno-mathematics, inferring that there is a cultural component to mathematics. I believed that prior to my exposure to ethno-mathematics that mathematics was acultural. The problem with this conceptualization of mathematics is that it suggests that a student's poor performance is based on a lack of innate ability. Parallel to this idea, the notion of mathematical ability being linked to genetics can be illustrated when a White professor walks into a classroom that includes racial and ethnic minority students, and there is an unspoken notion that they will probably not be successful. I held onto my acultural beliefs about mathematics until two things happened: the first was an encounter with my advisor in the mathematics department; the second was with one of my students. The professor who advised me throughout my master's degree and became my thesis adviser was known throughout the department for his work with students. The professor had successfully shepherded more students through their master's and doctoral degrees than any other professor in the department. As a student, I experienced the professor as extremely helpful. My studies with the faculty member did not generally involve discussions of pedagogy in a mathematics classroom as the professor was an Algebraist and not a mathematics educator. But, one day I explored this topic with the professor and asked, "Why do you think that the racial and ethnic minority students in the mathematics classes I am teaching at the community college are having such a difficult time passing my class?" The professor responded, "Well, it's genetic isn't it?" Talk about shock; I was surprised by the response. Was the professor really suggesting that there was some sort of math gene? Moreover, was the professor suggesting that racial and ethnic minority students did not possess it?

The second key experience that challenged me to think about the notion that mathematics constituted an innate element involved a student in my class, who reinforced the need for me to think about issues of cultural relevant pedagogy. I was speaking to one of my minority students during office hours about his struggles in the mathematics course. We were going over some problems and the student stopped, looked at me and said, "I can't do this math. Math is for White people." Being a Black woman who had just recently completed a master's degree in mathematics, I had to pause for a second and reflect on what this student was saying. "What do you mean?" I asked him. He replied, "It's got nothing to do with me. It doesn't make any sense." He reminded me of a film

I have seen entitled *Stand and Deliver*. There was a scene in the film where the teacher tried to make math relevant to his Mexican American students by telling them that the number zero was a concept created by the Mayans who were their ancestors. In the film, the mathematics instructor was responding to a student who asserted that "Math is for White people." My student and the student in the film were responding to that fact that they did not see their culture reflected in the activities in a math class. Interestingly, Ferguson (2003) and Martin (2006) suggest that teachers generally have low expectations of minority students and do not expect them to do well in mathematics. I recognized that I had been complicit in constructing the very same learning environment for my students.

I begin to employ three strategies that would turn around my classroom. The first involved writing and recognizing that students were finding it difficult to take notes in class. As a result, I provided online notes that the students could bring to class which highlighted the topics that we would be covering in class. This strategy was a significant support for many of my students. The second involved creating a more collaborative learning community in the classroom. When I used to describe my classroom set up as being rows of soldiers marching to destroy the enemy (me), it almost felt hostile. My class looked like the typical mathematics class in many colleges; the seats were lined up in straight rows with the students' attention being directed toward the front of the class where I would lecture to them. The set-up was perfect in the knowledge-transmission paradigm I was using as a mode of instruction. Borrowing from the work being done in K-12 mathematics classes, I changed my class and decided I would arrange my room in groups. In a quest to make mathematics have meaning, particularly for the racial and ethnic minority students, I also employed the concepts of ethnomathematics and culturally relevant teaching practices.

Changing Pedagogy: Culturally Relevant Teaching Practices—Ethnomathematics

Ethnomathematics (D'Ambrosio, 1985; Powell & Frankenstein, 1997) promotes the idea that as students learn mathematics they should incorporate their lived experiences and unique perspectives in a way that gives mathematics meaning to them. An ethnomathematics approach seeks to encourage students to see potential for mathematics in their lives outside of the classroom. Ethnomathematics infused in the curriculum helps to shatter the myth that "Math is for White people" by connecting mathematical knowledge to the student's own experience and by introducing a non-Eurocentric historical perspective of mathematics, which highlights the contributions of other cultures to mathematics development. Ethnomathematics has been found to provide a viable framework for a culture-inclusive mathematics curriculum. Anderson (1990) showed that introducing ethnomathematical perspectives into a curriculum enabled racial and ethnic minority students to develop a more self-confident attitude toward mathematics. In my own classroom, I used the example based in the research of Gilmer and Porter (1998), who suggest that that the first step in understanding ethnomathematics is going into a community and examining the communities experience with mathematical ideas. Using the geometrical designs commonly used in Black communities (e.g., hair braiding and weaving practices), Gilmer brought together the intricate patterns made by hair stylists and the mathematical concept of tessellations. Using the practical activity of hair braiding, Gilmer helped students to examine an application of mathematics that was relevant to them.

Borrowing from Gilmer and Porter (1998), I brought models into the classroom with their hair braided in different designs; we did development cost models based on the design to be used to project the cost of braiding one's hair. We also looked at determining the percentage of braids covering a model's head and examined the patterns of the braids which introduced the class to the mathematical concept of tessellations. At different points in the semester, I would take a few moments to talk about different counting systems, which allowed for a discussion of the Mayan concept of zero. I have had positive feedback from the classes in which I used ethnomathematics techniques; the more common being that the class was an enriching learning experience. Prior to incorporating these techniques, I can say the word enriching is not one I would normally hear in a mathematics class. I integrated ethnomathematics into my classroom as a strategy to challenge the Western and Eurocentric framework embedded in educational processes. Moreover, I experienced that one of the key impacts on my students came when I changed from individual to a collaborative group.

Bringing Culturally Relevant Approaches to Mathematics Pedagogy to Community College

Ladson-Billings (1994) created the term *culturally relevant pedagogy* to describe "a pedagogy that empowers students intellectually, socially, and politically by using cultural references to impart knowledge, skills, and attitudes" (pp. 17–18). According to Ladson-Billings (1994), teachers integrating this pedagogical and methodological work into the classroom promoted strong connections between a student's home and school life by utilizing the students' backgrounds, knowledge, and experiences to inform the teachers' class instruction. A number of researchers (Gay, 2000; Ladson-Billings, 1994, 1995, 1997; Tate, 1995) have contributed to development of practices involved in culturally relevant pedagogy. A fundamental belief of these practitioners is that "all of their students can succeed rather than that failure is inevitable for some" (Ladson-Billings, 1994, p. 25). Ladson-Billings (1995) suggests that culturally relevant pedagogy has its basis in three main propositions: (a) Students must experience academic success; (b) Students must maintain and/or develop cultural competence; and (c) Students must develop a critical consciousness.

Culturally Relevant Pedagogy in My Courses

Using a culturally relevant approach in my classroom meant that I had to change my pedagogical approach in teaching. Instead of using a traditional lecture approach to teaching mathematics, I used a collaborative approach as much as possible. According to Ladson-Billings (1995, 1997), a collaborative, cooperative learning community is one in which the teacher and students are predisposed to the idea that everyone must succeed as opposed to the competition model in a traditional mathematics positions students as individuals acting only in their best interest. Research has shown that collaborative learning communities have been shown to have a strong impact on the performance of Black students and other racial and ethnic minority students (Tate, 1995) in the classroom. A key approach in my courses was to work collectively toward a common goal of getting students to invest in each other's achievement. Based on the group work that I assigned in and out of class, emphasis was placed on the group understanding the work not just the individual. I would assign groups based on a student's ability so that I would have stronger and weaker students working together. To make sure the groups were operating cooperatively, I would work with the groups individually, encouraging stronger members to help the weaker members. Our class goal was for everyone to learn, not just a fortunate few. In the group work, I worked to create mathematical problems that resonated with the group and had meaning for them. When students are engaging their learning processes, they can be extremely creative. One semester, I had a student who came up with a rap song to explain the five stages of solving an application (or the dreaded word problem). It was hilarious, the class loved it, and for many it made connecting to the process more understandable.

Working toward a class that bought into the idea of cooperative learning took some work, since most students were happy to experience the mathematics course in the traditional manner where I behaved as the transmitter of knowledge and they behaved as the receptacles of that knowledge (Freire, 1971). The effort was well worth it because, over time, I saw my 70% failure rate reduced to less than 20%. My experiences have taught me that there is a need to elevate the discussion of pedagogical approaches in a community college classroom so that instructors do not just defer to the traditional classroom practices. This is important because of the need to meet the forecasted STEM worker shortages. This cannot be done without an effort to effectively increase the number of racial and ethnic minority students who successfully complete STEM degrees. This goal cannot be achieved without removal of the mathematical barrier which serves as an impediment to racial and ethnic minority students to complete STEM degrees.

Conclusion

Why is this discussion important? There are three important reasons. The first is there is a need for increasing the number of students who can successfully complete STEM degrees and enter STEM careers. The second relates to the demographic shift in our society, suggesting that a greater number of these jobs will have to be filled by racial and ethnic minorities who have historically been underrepresented in STEM fields (e.g., Blacks, Latinos/as, and Native Americans). Third, the students from these underrepresented groups have been impeded in the quest for a STEM degree by the gatekeeper course of mathematics.

Mathematical pedagogical practices in community colleges will need to change to make mathematics more accessible to racial and ethnic minority students. As discussed, using culturally relevant approaches coupled with ethnomathematics allowed minority students to perform better in mathematics in my classes. Since educators know mathematics is a gatekeeper to STEM, particularly for minority students, teaching mathematics based on a Eurocentric or Western framework and employing only traditional methods will not be acceptable. Community college faculty must make mathematics something that students learn together. Furthermore, community college educators must make the effort to integrate math into the lives of their students by culturally connecting the concepts to them. Community college professors must be more strategic in their pedagogical methods to develop and implement culturally relevant learning communities in the classroom that promote the academic achievement of racial and ethnic minority students.

References

Abrahamson, J. A., & Cooper, H. F. (1993). *What did we get for our $30 billion investment in SDI/BMD?* Retrieved from http://www.nipp.org/National%20Institute%20Press/Archives/Publication%20Archive%20PDF/What%20for%20$30B_.pdf

Anderson, S. (1990). Worldmath curriculum: Fighting eurocentrism in mathematics. *Journal of Negro Education, 59*(3), 348–359.

Ashcraft, M. H., & Kirk, E. P. (2001). The relationships among working memory, math anxiety, and performance. *Journal of Experimental Psychology: General, 130*(2), 224–237.

Byrd, K. L., & Macdonald, G. (2005). Defining college readiness from the inside out: First-generation college student perspectives. *Community College Review, 33*(1), 22–30.

Carnevale, A., Smith, N., & Strohl, J. (2010). *Help wanted: Projections of jobs and education requirements through 2018.* Retrieved from http://cew.georgetown.edu/jobs2018

Crisp, G., Nora, A., & Taggart, A. (2009). Student characteristics, pre-college, college and environmental factors as predictors of majoring in and earning a STEM degree: An analysis of students attending a Hispanic serving institution. *American Educational Research Journal, 46*(4), 924–942.

D'Ambrosio, U. (1985). Ethnomathematics and its place in the history and pedagogy of mathematics. *For the Learning of Mathematics, 5*(1), 44–48.

Ferguson, R. (2003). Teachers' perceptions and expectations and the black-white test score gap. *Urban Education, 38*(4), 460–507.

Freire, P. (1971). *Pedagogy of the oppressed.* New York, NY: Continuum.

Gasiewski, J. A., Eagan, M. K., Garcia, G. A., Hurtado, S., & Chang, M. J. (2011). From gatekeeping to engagement: A multicontextual, mixed method study of student academic engagement in introductory STEM courses? *Research in Higher Education, 53*(2), 229–261.

Gay, G. (2000). *Culturally responsive teaching.* New York, NY: Teachers College Press.

Gilmer, G., & Porter, M. (1998). Hairstyles talk a hit at NCTM. *International Study Group on Ethnomathematics Newsletter, 13*(2), 5–6.

Hagedorn, L. S., & Purnamasari, A. V. (2012). A realistic look at STEM and the role of the community colleges. *Community College Review, 40*(2), 145–164.

Herrnstein, R., & Murray, C. (1994). *The bell curve.* New York, NY: The Free Press.

Hoyt, J. E. (1999). Developmental education and student attrition. *Community College Review, 27*(2), 51–72.

Kirsch, I., Braun, H., & Yamamoto, K. (2007). *America's perfect storm: Three forces changing our nation's future.* Princeton, NJ: Policy Evaluation and Research Center, Policy Information Center, Educational Testing Service.

Ladson-Billings, G. (1994). *The dreamkeepers: Successful teaching for African-American students.* San Francisco, CA: Jossey-Bass.

Ladson-Billings, G. (1995). But that's just good teaching! The case for culturally relevant pedagogy. *Theory into Practice, 34*(3), 159–165.

Ladson-Billings, G. (1997). It doesn't add up: African American students' mathematics achievement. *Journal for Research in Mathematics Education, 28*(6), 697–708.

Lowell, B. L., & Salzman, H. (2007). Into the eye of the storm: Assessing the evidence on science and engineering education, quality, and workforce demand. Retrieved from http://www.urban.org/UploadedPDF/411562_salzman_Science.pdf

Lucore, R. (2012). Community colleges an unsung source. Retrieved from http://www.huffingtonpost.com/rebecca-lucore/community-colleges-an-uns_b_1861830.html

Maltese, A. V., & Tai, R. H. (2011). Pipeline Persistence: Examining the association of educational experiences with earned degrees in STEM among U.S. Students. *Science Education, 95*(5), 877–907.

Martin, D. (2006). Mathematics learning and participation as racialized forms of experience: African American parents speak on the struggle for mathematical literacy. *Mathematical Thinking and Learning, 8*(3), 197–229.

Martin, D. (2012). Learning mathematics while Black. *The Journal of Educational Foundations, 26*(1-2), 47–66.

McCabe, R. (2003). *Yes, we can: A community college guide for developing America's underprepared.* Phoenix, AZ: League for Innovation in the Community College.

Mesa, V., Celis, S., & Lande, E. (2011). Teaching approaches of community college mathematics faculty: Do they relate to classroom practices? Retrieved from http://141.213.232.243/bitstream/2027.42/85214/1/MesaCelisLande_2011_DeepBluePost.pdf

Moses, R. P., & Cobb, C.E. (2001). *Radical equations.* Boston, MA: Beacon Press.

Mullin, C. M. (2011, October). *The road ahead: A look at trends in educational attainment by community college students* (Policy Brief2011-04PBL). Washington, DC: American Association of Community Colleges.

Obama, B. (2006). *The audacity of hope: Thoughts on reclaiming the American dream.* New York, NY: Vintage Books.

Obama, B. (2009, July 14). Remarks by the president on the American Graduation Initiative. Retrieved from http://www.whitehouse.gov/the_press_office/Remarks-by-the-President-on-the-American-Graduation-Initiative-in-Warren-MI

Obama, B. (2010). Remarks by the president in State of Union Address. Retrieved from http://www.revistalafactoria.eu/articulos/VO%20AMERICANA.pdf

Öztürk, M. D. (2007). Global competition: American's underrepresented minorities will be left behind. Retrieved from http//www.tcrecord.org

Palmer, R. T., Maramba, D., & Dancy, T. (2011). A qualitative investigation of factors promoting the retention and persistence of racial and ethnic minority students in STEM. *The Journal of Negro Education, 80*(4), 491–504.

Pendergast, A. (1989). *Everybody counts: A report to the nation on the future of mathematics education*. Washington, DC: National Academy Press.

Powell, A., & Frankenstein, M. (1997). *Ethnomathematics: Challenging eurocentrism in mathematics education*. Albany: State University of New York Press.

Ryan, C. L., & Siebens, J. (2012). *Educational attainment in the United States: 2009 (current population reports)*. Retrieved from http://www.census.gov/prod/2012pubs/p20-566.pdf

Tate, W. F. (1995). Returning to the root: A culturally relevant approach to mathematics pedagogy. *Theory into Practice, 34*(3), 166–173.

Thernstrom, S., & Thernstrom, A. (1997). *America in Black and White: One nation indivisible*. New York, NY: Simon & Schuster.

Thernstrom, A., & Thernstrom, S. (2004). *No excuses: Closing the racial gap in learning*. New York, NY: Simon & Schuster.

Uttal, D. H. (1996). Beliefs, motivation, and achievement in mathematics: A cross-national perspective. In M. Carr (Ed.), *Motivation in mathematics* (pp. 25–37). Cresskill, NJ: Hampton.

PART III

Examining the Experiences of Minority Students in Community Colleges

Diverse Contexts

9

MINORITY SERVING COMMUNITY COLLEGES AND THE PRODUCTION OF STEM ASSOCIATE'S DEGREES

Frances King Stage, Ginelle John, Valerie C. Lundy-Wagner, and Katherine Mary Conway

The disparity between growing postsecondary enrollments of minority students and their persistently low numbers in the science, technology, engineering, and mathematics (STEM) fields puts the role of the United States as a global leader at risk. Minority student enrollments in STEM have not kept pace with minority participation in higher education overall, exacerbating a growing deficiency of STEM graduates. Minority college enrollment doubled between 1976 and 2007 from 15% to 32% (U.S. Department of Education, 2009), and the number of degrees awarded to these groups increased, but not across all disciplines, institution types, or for all racial/ethnic groups.

The scarcity of students selecting a STEM major has resulted in initiatives to generate awareness of, access to, and success in these fields, with an emphasis on minorities in minority-serving institutions (MSIs; NAS, 2010). The disproportionate enrollment of Asian, African American, Hispanic, and Native American students at community colleges (Baum & Ma, 2007) is another important opportunity for promoting STEM. However, despite the emphasis on MSIs that enroll high proportions of minority students in STEM at the baccalaureate level, relatively little research exists on MSI community colleges and their contribution to the minority STEM pipeline. This lack of work is problematic for three reasons: (a) the success of four-year MSIs in producing minority STEM degree graduates, (b) community colleges serve as an important entry point for many minority students, and (c) the recent decrease in STEM degrees awarded at community colleges. We know little about the contribution of two-year institutions.

MSIs are important for underrepresented minority STEM participation and success. While many are familiar with historically Black colleges and universities (HBCUs) and their success in STEM (Gasman, Lundy-Wagner, Ransom,

& Bowman, 2010; Hubbard & Stage, 2009; Perna et al., 2009), less is known about other MSIs (e.g., Tribal Colleges and Universities (TCUs), Hispanic-serving institutions (HSIs), and Asian American and Native American Pacific Islander-serving institutions (AANAPISIs) and their contributions to minority STEM degree receipt. Here we add to the literature by describing the contribution of two-year MSIs to minority receipt of STEM associate's degrees and sub-baccalaureate certificates. Additionally, we contextualize the analysis by highlighting STEM and non-STEM completions at two-year colleges by racial/ethnic group at MSIs and non-MSIs. The chapter concludes by describing policy implications related to the importance of two-year MSIs in promoting STEM majors; we also discuss ways two-year MSIs can contribute to additional STEM achievement.

The purpose of this chapter is to add to existing research related to MSI contributions toward the minority STEM pipeline by including contributions of two-year MSIs on minority receipt of STEM associate's degrees and sub-baccalaureate certificates. Recent studies have shown that Latino students who complete associate's degrees are more likely to complete four-year degrees (Morales, 2011; Roksa & Calcagno, 2008). MSIs play a pivotal role in U.S. higher education, serving approximately 35% of the U.S. minority population, although accounting for less than 3% of all college enrollments. Additionally, 55% of minorities enrolled at MSIs are in public two-year colleges (Li, 2007). We highlight STEM and non-STEM completions at two-year colleges by racial/ethnic group at MSIs and non-MSIs.

Background

Minority-Serving Institutions

MSIs fall into two categories: those legislated, and those whose status is a function of enrollments of minority students. The former includes HBCUs and TCUs identified in Title III of the Higher Education Act of 1965, Equity in Educational Land-Grant Status Act of 1994, Tribally Controlled Community College Assistance Act of 1978, or Navajo Community College Assistance Act of 1978. The second category includes institutions whose percentages of minority students have led to their designation as an MSI. The Higher Education Act of 1965 gave special MSI status to postsecondary schools whose enrollment of a single minority or a combination of minorities exceeded 50% of the total enrollment. The term *minority* means American Indian, Alaskan Native, Black (not of Hispanic origin), Hispanic (including persons of Mexican, Puerto Rican, Cuban, and Central or South American origin), Pacific Islander or other ethnic group (Higher Education Act of 1965). HBCUs are designated based on year established (prior to 1964) (Gasman et al., 2010). TCUs must have 50% Native enrollment (Gasman, Baez, & Turner, 2008). HSIs must have 25%

Hispanic full-time equivalent enrollment, and 50% must be low income (Contreras, Malcom, & Bensimon, 2008). AANAPISIs must have 10% AANAPISI enrollment and 50% must be low-income.

MSIs represent a heterogeneous set of institutions; most with missions (implicit or explicit) to serve historically underrepresented and/or economically disadvantaged students. MSIs are especially important for providing college access given that they are three times more likely to have open admissions policies than non-MSIs, a feature that highlights their commitment to providing all students with postsecondary access (Li, 2007).

Research has shown that MSIs have a positive impact on minority student success in STEM (Fries-Britt, Younger, & Hall, 2010b; Kim & Conrad, 2006; National Science Board, 2004; Palmer, Davis, & Thompson, 2010; Solórzano, 1995; Stage, John, & Hubbard, 2011; Stage, Lundy-Wagner, & John, 2012; Suitts, 2003). One reason for that success is the validation students receive from peers of the same racial or ethnic background (Museus, Jayakumar, & Robinson, 2011) in a supportive environment that includes role models (Allen, 1992; Lent et al., 2005; Palmer, 2010; Palmer & Gasman, 2008; Hurtado, Alvarez, Guillermo-Wann, Cuellar, & Arellano, 2012). A hostile or chilly campus climate that incorporates who does science, what they look like, and how they should act has been shown to discourage minority students in STEM at both two- and four-year institutions (Bensimon, 2003; Fries-Britt, Younger, & Hall, 2010a; Maple & Stage, 1991; Oakes, 1990) and leads to STEM program drop-out (Bonous-Hammarth, 2000; Hurtado et al., 2007; Leslie, McClure, & Oaxaca, 1998). A study of STEM students at an HBCU found that students benefited from smaller class sizes, peer interactions, faculty accessibility and academic support services such as tutoring (Perna et al., 2009), experiences that were often lacking for minority students at predominantly White institutions (Museus, Palmer, Davis, & Maramba, 2011).

However, research showing that MSIs positively contribute to minority student success is mitigated by other research suggesting that beginning at an open enrollment community college can hinder baccalaureate attainment. One study, using national data, employed linear regression techniques and concluded that attendance at an MSI was positively and significantly correlated with Black, Hispanic, and Native American students' academic success in STEM. However, when the interaction effect of institutional selectivity and MSI status was examined, the positive benefit of attending an MSI disappeared (Hurtado et al., 2007).

The proportion of minority students in STEM contrasts with proportions of minority students enrolled generally, many of whom begin at the community colleges (NAS, 2010). Community colleges enroll 44% of undergraduates: 43% of first-time freshmen; 55% of Native American students; 52% of Hispanic students; 45% of Asian/Pacific Islander students, and 44% of Black students (American Association of Community Colleges, 2011; Shannon &

Smith, 2006). Despite their representation, minorities at two-year institutions are less likely to major in or complete a STEM degree (Expanding Underrepresented Minority Participation, 2011). One reason may be a lack of preparation: community colleges often have open admissions policies, accepting students who have not taken a college preparatory curriculum and who need remedial coursework (National Center for Education Statistics, 2011). This is problematic for minority STEM participation since high school performance (e.g., GPA, math and science course taking, standardized test scores) is positively correlated with majoring in STEM (Maple & Stage, 1991; Ware & Lee, 1988). However, most research describing or predicting minority STEM participation is based on bachelor's degree-seeking student attitudes, behaviors, and/or achievement.

For students who begin at community colleges, estimates of successful transfer range from 25% to 52% (Hoachlander, Sikora, Horn, & Carroll, 2003). Accurate data is difficult to attain because community college students often attend part time, transfer outside their university system, and are difficult to track (Bailey et al., 2004; Chen & Weko, 2009). Little research exists on STEM transfer students, but transfer success in STEM is even more problematic and students in the sciences are often told that only one in three will ultimately graduate (Goodchild, 2004; Kinzie, 2002).

STEM Initiatives in Community Colleges

Despite poor minority participation in STEM at community colleges, there is growing awareness of these colleges' capacity for increasing the STEM pipeline. This has resulted in various initiatives, many funded by the National Science Foundation (NSF). For example, the Louis Stokes Alliances for Minority Participation Program was designed to increase the number and quality of underrepresented students earning STEM undergraduate degrees/certificates and to expand the number interested in and qualified for graduate programs. Within this program many awards went to projects focused on doctoral education, but recent initiatives have allocated funds to community colleges (Mooney & Foley, 2011).

Community colleges often partner with baccalaureate institutions to engage students earlier, create science awareness, offer courses to aid transfer, develop research opportunities and provide mentors to improve STEM participation. According to Tsui (2007), these efforts are effective for increasing minority STEM participation; they incorporate holistic support including financial aid and peer support that build on Tinto's (1975) notions of academic and social integration (see Chang, Cerna, Han, & Saenz, 2008). Financial remuneration (stipend or summer employment) is essential for community college students for whom money is often a factor in attrition (Dowd & Coury, 2006; Hagedorn, Maxwell, & Hampton, 2002; King, 2002; Paulsen & St. John, 2002) and for MSIs in particular, because they are more likely to enroll low-income students (Li, 2007). Minority student attrition in community college STEM programs is related to insufficient financial resources (Hurtado et al., 2007; Maton & Hrabrowski, 2004; May & Chubin, 2003; Perna et al., 2009) including STEM scholarship awards that are merit based, disadvantaging underprepared minority students and further limiting their participation (Fenske, Porter, & DuBrock, 2000). Finally, studies suggest that the likelihood of students both majoring in science and pursuing an advanced degree increase if students participate in undergraduate research (Barlow & Villarejo, 2004; Hurtado et al., 2007; Lopatto, 2004).

While funded programs are successful in attracting community colleges students to STEM, they affect a small percentage of MSIs. Additionally, most initiatives are too new to identify long-term benefits and effectiveness is difficult to assess since most students attend part-time, disqualifying them for inclusion in persistence or completion calculations (Bailey & Alfonso, 2005; Bailey, Crosta, & Jenkins, 2006). Finally, to date few, if any, studies have classified community colleges by their MSI status and attempted to assess their role in minority STEM associate's or sub-baccalaureate certificate attainment.

This research takes a critical quantitative perspective (Stage, 2007). Critical quantitative work uses numerical data to challenge the status quo and to reveal negative assumptions and contradictions within the traditional quantitative analysis. By disaggregating data by racial/ethnic groups, we can see problems as well as successes in STEM pathways. Results from this study will inform subsequent research analyzing success for two-year transfer students who seek Bachelor's degrees in STEM as well as research seeking to identify optimal college environments for diverse STEM students.

Method

To identify the contribution of two-year MSIs on the production of minority STEM associate's and sub-baccalaureate certificates, we used a two-tiered approach. First, we reviewed extant literature focusing on three substantive areas: MSIs in higher education, STEM at community colleges, and STEM initiatives beyond the classroom in community colleges. Second, we conducted a descriptive analysis of a cohort of community college student associate's and sub-baccalaureate certificate awardees using Integrated Postsecondary Education Data System (IPEDS, 2010).

Data

IPEDS consists of annual surveys collected by the National Center of Education Statistics (NCES). These surveys include data (i.e., institutional characteristics, enrollment and student characteristics, persistence, tuition, financial aid, completion, finance, and staffing) from all postsecondary institutions receiving federal student aid. We used the Institutional Characteristics data to identify institutions awarding associate's degrees and sub-baccalaureate certificates that were also designated MSIs (AANAPISIs, HSIs, HBCUs, and TCUs) for 2008-2009. Using Cohen and Brawer's (2008) definition of community college, we included institutions where bachelor's degrees accounted for fewer than 10% of all undergraduate degrees.

Completion data includes information on the number of awards and degrees conferred in a given year at an institution by academic program, award level, race/ethnicity, and gender. Using this data we identified the production of STEM associate's degrees and sub-baccalaureate certificates (STEM included mathematics/statistics, agricultural, biological and physical sciences, engineering/engineering technologies, and computer/information sciences), providing a snapshot of the two-year MSI contribution to minority STEM participation.

Results

Table 9.1 displays data by race/ethnicity and institutional type (MSI vs. non-MSI) for STEM associate's degrees and certificates. Results show that 1,730 postsecondary institutions granted associate's and sub-baccalaureate certificates in 2008-2009 (although not all 1,730 had STEM associate's degree or STEM certificate programs). Altogether 26,694 STEM certificates and 53,602 STEM associate's degrees combined for a total of slightly more than 80,000 awards conferred by two-year colleges. American Indian, Asian, Black, and Latino students earned 15,566 associate's degrees in STEM, compared with 34,601 White students. The remaining degrees were awarded to students classified as non-residents (962) and other (2,473).

Table 9.1 compares STEM associate's degrees and sub-baccalaureate certificate completions awarded in MSIs and non-MSIs. The percentage of STEM associate's degrees earned at MSIs compared to the total associate's STEM degrees earned ranged from 11.5% for Asian students to just 1.3% for White students. For Native American students, 10% of all STEM associate's degrees were earned at an MSI. For Hispanic/Latino students 5% and for Black students 4% of all STEM associate's degrees were earned at an MSI community college.

White students earned 66% of total STEM associate's degrees awarded at non-MSIs and 31% at MSIs. On the other hand, Asian students earned 21% of STEM associate's degrees at MSIs and just 5% of STEM associate's degrees at non-MSIs. Similarly, Latinos earned 18% of STEM associate's degrees at MSIs and 11% of the degrees at non-MSIs. For American Indian/Alaskan Native students the proportions were 4% at MSIs and 1% at non-MSIs, and for Blacks the similar comparison was 14% and 11%, respectively.

By contrast, Table 9.2 presents non-STEM associate's degrees awarded during the same period. The data suggest that receipt of STEM associate's degrees and sub-baccalaureate certificates mirrors attainment by race, with some

TABLE 9.1 Number of STEM Associate's Degrees and Certificates Awarded by Race July 1, 2008 to June 30, 2009

Community Colleges (N = 1,730)*	American Indian or Alaskan Native		Asian/Native Hawaiian/ Other Pacific Islander		Black or African American		Hispanic or Latino		White		Non-Residents		Other		Total
	#	%	#	%	#	%	#	%	#	%	#	%	#	%	
STEM															
MSIs Awarding Associates n = 67/71	58	3.8	323	21.0	219	14.2	276	18.0	477	31.0	104	6.8	80	5.2	1,537
MSIs Awarding Certificates n = 45/71	63	4.9	225	17.6	115	8.9	367	28.6	422	32.9	18	1.4	72	5.6	1,282
Non-MSIs Awarding Associates n = 1,190/1,243	512	1.0	2,476	4.8	5,959	11.4	5,743	11.0	34,124	65.5	858	1.6	2,393	4.6	52,065
Non-MSIs Awarding Certificates n = 852/1,243	256	1.0	1,117	4.4	3,334	13.1	3,472	13.7	15,587	61.3	261	1.0	1,385	5.5	25,412
Totals															
Associates	570	1.1	2,799	5.2	6,178	11.5	6,019	11.2	34,601	64.0	962	1.8	2,473	4.6	53,602
Certificates	319	1.2	1,342	5.0	3,449	12.9	3,839	14.4	16,009	60.0	279	1.0	1,457	5.5	26,694

* IPEDS Carnegie Classification 2005: Basic- Associate's Colleges and Degree Granting Associates and Certificates. Data was obtained from the National Center for Education Statistics' Integrated Postsecondary Education Data System (IPEDS). Data Source: Completions- Awards in science, technology; engineering, and mathematics (STEM) July 1, 2008 to June 30, 2009.

Note: STEM majors include: mathematics/statistics; natural sciences (agricultural, biological and physical sciences); engineering/engineering technologies; and computer/information sciences

TABLE 9.2 Number of Non-STEM Associate's Degrees and Certificates Awarded by Race July 1, 2008 to June 30, 2009

Community Colleges (N = 1,730)*	American Indian or Alaskan Native		Asian/Native Hawaiian/Other Pacific Islander		Black or African American		Hispanic or Latino		White		Non-Residents		Other		Total
	#	%	#	%	#	%	#	%	#	%	#	%	#	%	
Non-STEM															
MSIs Awarding Associates n = 101/103	1,316	4.1	4,658	14.6	4,336	13.6	8,198	25.7	9,731	30.5	1,385	4.3	2,311	7.2	31,935
MSIs Awarding Certificates n = 68/103	465	2.7	2,473	14.2	2,021	11.6	5,446	31.3	5,349	30.7	352	2.0	1,309	7.5	17,415
Non-MSIs Awarding Associates n = 1,609/1,651	5,275	0.9	27,326	4.7	66,267	11.4	67,932	11.7	370,648	64.0	10,615	1.8	30,757	5.3	57,8820
Non-MSIs Awarding Certificates n = 1,431/1,651	4,357	1.1	14,089	3.6	68,880	17.7	48,537	12.4	229,699	58.9	2,726	0.7	21,596	5.5	389,884
Totals															
Associates	6,591	1.1	31,984	5.2	70,603	11.6	76,130	12.5	38,0379	62.3	12,000	2.0	33,068	5.4	61,0755
Certificates	4,822	1.2	16,562	4.1	70,901	17.4	53,983	13.3	235,048	57.7	3,078	0.8	22,905	5.6	407,299

* IPEDS Carnegie Classification 2005: Basic–Associate's Colleges and Degree Granting: Associates and Certificates. Data was obtained from the National Center for Education Statistics' Integrated Postsecondary Education Data System (IPEDS). Data Source: Completions– Awards in science, technology, engineering, and mathematics (STEM) July 1, 2008 to June 30, 2009.

Note: STEM majors include: mathematics/statistics; natural sciences (agricultural, biological and physical sciences); engineering/engineering technologies; and computer/information sciences exceptions. For example, Asian students earned a similar proportion of STEM and non-STEM associate's degrees at MSIs and non-MSIs (about 5%); however, they were conferred 21% of STEM degrees by MSIs compared to only 15% of non-STEM degrees by MSIs. Differing in proportions from Asians, Hispanic/Latino students earned about 26% of non-STEM associate's degrees and a lower percentage, 18% of STEM degrees at MSIs. Native American students are underrepresented in general in both two- and four-year institutions. Yet they earned 4% of STEM and non-STEM associate's degrees awarded by MSI's. This contrasts with similar data for four-year institutions where American Indian students earned 0.5% of STEM degrees awarded by MSI institutions and 1% of STEM degrees awarded by non-MSIs (Stage et al., 2012).

Table 9.3 presents the number and percentage by race of STEM degrees awarded by 67 two-year MSIs compared with numbers by race awarded by 1,190 non-MSIs. Overall 103 MSIs represent less than 6% of two-year colleges awarding associate's degrees and awarded 6% of the STEM associate's degrees earned by underrepresented minority students. Two-year MSIs awarded percentages of STEM associate's degrees ranging from 1% for White students to 12% for APA students. Additionally they award 10% of STEM associate's degrees for American Indian students. These numbers are more impressive when we consider the average enrollment size of MSIs compared with non-MSIs.

In general, two-year MSIs tend to have lower student enrollments compared to non-MSIs, yet they serve American Indian and APA students well in terms of the percentage of associate's degrees earned. On the other hand, Black and Latino students earned associate's degrees in relatively lower proportions. Only 5.3% (or 67) of the 1,257 institutions granting STEM minority completions were MSIs, but they contributed 6% of STEM associate's degrees, and 9% of

TABLE 9.3 Number and Percentage of STEM Associate's Degrees at MSIs and Non-MSIs by Race

	# STEM Assoc Total	# STEM Assoc (MSI)	% STEM Assoc (MSI)	# STEM Assoc (Non-MSI)	% STEM Assoc (Non-MSI)
American Indian or Alaskan Native	570	58	10	512	90
Asian/Native Hawaiian/Other Pacific Islander	2,799	323	12	2,476	88
Black or African American	6,178	219	4	5,959	96
Hispanic or Latino	6,019	276	5	5,743	95
White	34,601	477	1	34,124	99
Non-Residents	962	104	11	858	89
Other	2,473	80	3	2,393	97

TABLE 9.4 STEM Associate Degrees and Certificates Awarded to Underrepresented Minority Students July 2008 to June 2009.

	At MSIs	%	At non MSIs	%
Total Associate Degrees Awarded	19,384	9.6	181,490	90.4
STEM Associate Degrees Awarded	876	5.6	14,690	94.4
STEM degrees as a proportion of total degrees awarded		4.5		8.1
Total Certificates Awarded	11,175	7.2	144,042	92.8
STEM Certificates Awarded	770	8.6	8,179	91.4

sub-baccalaureate certificates. Finally, Table 9.4 summarizes the absolute numbers and percentages of STEM associate's degrees and certificates awarded to minority students. Minority students earned approximately 10% of the associate's degrees awarded but only 6% of the total STEM associate's degrees at the same institutions.

Stage et al. (2012) found that four-year MSIs awarded a greater proportion of STEM degrees than their representation overall. By contrast, students at two-year MSIs performed better in the earning of STEM certificates. Minority students were awarded 7% of all certificates and 8.6% of STEM certificates. At first, these numbers might suggest that minority students fare worse by attending MSIs. However, the two-year MSIs have low enrollments compared with their non-MSI counterparts, tend to have a smaller range of possible majors, and tend to operate with limited budgets, including science facilities that with limited resources. Even with relatively lower percentages of STEM majors, these institutions serve as a resource and an opportunity for student of color.

Discussion

In general, there is a lack of research focusing on MSI community colleges. This is problematic for three reasons. First, gaps persist in postsecondary access and success in STEM fields (NSF, 2009). Given that two-year MSIs disproportionately enroll and graduate marginalized students in associate's degree/certificate programs, education researchers should continue to examine these institutions and explore ways to expand STEM opportunities. Latino students who complete associate's degrees are more likely to complete four-year degrees (Morales, 2011; Roksa & Calcagno, 2008). Second, MSIs often provide academically under-prepared students with pathways into STEM. While calls to increase the pipeline often include attention to ethnicity/race (NAS, 2010), others have described the institutional context as an equally important consideration (Dowd, 2003; Perna et al., 2009). Further, as K-12 and remedial education suffers through funding cuts, the most vulnerable students, many of whom will likely attend an MSI if they choose postsecondary education will be most affected. Third, MSIs have a history of success in preparing minorities and low-income students in STEM, but the focus is usually on HBCUs or HSIs (Gloria, Castellanos, Lopez, & Rosales, 2005). AANAPISIs and TCUs also play an important role in diversifying the undergraduate STEM pipeline. Research reported here focused on a national view of the role, in the aggregate, that minority-serving institutions play in the success of the undergraduate stem majors.

Recent work focuses on specific types of minority serving institutions and the role they play in underrepresented college student success (Bensimon, 2003; Contreras et al., 2008; Dowd, 2003; Palmer, Davis, & Hilton, 2009). More recently, a study of STEM productivity at four-year MSIs found that underrepresented students who enrolled in MSIs, in general, fared better than their counterparts at non-MSI institutions. Percentages of students from underrepresented minority groups earning STEM degrees at four-year MSIs by and large exceeded percentages of students from the same minority groups earning STEM degrees at non-MSIs (Stage et al., 2012).

Despite the small number of community colleges designated as MSIs, these institutions also can be important contributors to minority receipt of STEM associate's and sub-baccalaureate certificates. Not surprisingly, as in the previous study focusing on four-year institutions, the production of STEM completions at MSI community colleges varied by ethnicity/race. Nonetheless, in the aggregate these data highlight the importance of the two-year MSI in addition to the four-year MSI for expanding minority STEM participation. In addition to building on MSI STEM research, this work complements work focusing on STEM baccalaureate origins of underrepresented students (Stage & Hubbard, 2009; Wolf-Wendel, 1998), which found that specific types of MSIs are major contributors to the production of STEM scientists (Contreras et al., 2008).

This study builds upon past work by incorporating data from HBCUs and HSIs and also AANAPISIs and TCUs, which also play an important role in diversifying the undergraduate STEM pipeline. Percentages of students from underrepresented minority groups earning STEM degrees at four-year MSIs by and large exceed percentages of students from the same minority groups earning STEM degrees at non-MSIs. Results from this study are not as clear as for studies focusing on four-year MSIs. This study supports existing literature on MSIs and the production of underrepresented undergraduates receiving STEM degrees.

Findings suggest that the MSIs in this study not only produced a considerable number of minority STEM associate's degree recipients, but that STEM bachelor's degree production for designated MSIs and their respective ethnics/racial groups were important. Two-year MSIs served American Indian and APA students well in terms of the percentage of associate's degrees earned.

On the other hand, Black and Latino students earned associate's degrees in relatively lower proportions. Stage et al. (2012) found that many successful STEM students from several racial/ethnic groups attended four-year HBCUs. However, only 15 of 103 HBCUs are two-year colleges (Brown, 2003), reducing the relative influence of those particular institutions on the overall success underrepresented students in STEM majors.

Conclusion

Two-year MSIs represent a relatively small number of the thousands of postsecondary education institutions in the United States, yet they provide an environment for budding scholars, and particularly for underprepared students seeking STEM degrees. Given gaps in postsecondary access and success in STEM fields (NSF, 2009), two-year MSIs can provide academically under-prepared pathways into STEM. A question left unanswered is how many of these students who earn their associate's degree, transfer and go on to earn a Bachelor's degree or beyond in a STEM field. Research shows that two-year college students who earn the associate's degree are more likely to transfer and persist through baccalaureate degree completion. While it is difficult to track community college students as they move on to four-year institutions, it is possible to get estimates of their success by examining large systems such as New York City, California, and Florida. Finally, STEM collaborations between associate's and baccalaureate programs should be encouraged (Chang et al., 2008). Such cooperation can be important in mitigating problems transfer students experience adjusting to a more competitive and individualistic culture at a four-year college.

References

American Association of Community Colleges 2011 Fast Facts. (2011). Retrieved from http://www.aacc.nche.edu/AboutCC/Pages/fastfacts.aspx

Allen, W. R. (1992). The color of success: African-American college student outcomes at predominantly White and historically Black public colleges and universities. *Harvard Educational Review, 62*(1), 26–44.

Bailey, T., Alfonso, M., Calcagno, J. C., Jenkins, D., Kienzl, G., & Leinbach, T. (2004). *Improving student attainment in community colleges: Institutional characteristics and policies.* New York, NY: Community College Research Center, Institute on Education and the Economy, Teachers College, Columbia University.

Bailey, T. R., & Alfonso, M. (2005). *Paths to persistence: An analysis of research on program effectiveness at community colleges.* Indianapolis, IN: Lumina Foundation for Education.

Bailey, T. R., Crosta, P., & Jenkins, D. (2006, August). *What can student right-to-know graduation rates tell us about community college performance?* CCRC Working Paper. New York, NY: Columbia University, Teachers College, Community College Research Center.

Barlow, A. E. L., & Villarejo, M. (2004). Making a difference for minorities: Evaluation of an educational enrichment program. *Journal of Research in Science Teaching, 41*(9), 861–881.

Baum, S., & Ma, J. (2007). *Education pays: The benefits of higher education for individuals and society.* New York, NY: College Board.

Bensimon, E. M. (2003). The diversity scorecard: A learning approach to institutional change. *Change: The Magazine of Higher Learning, 36*(1), 44–52.

Berryman, S. E. (1983). *Who will do science?* New York, NY: Rockefeller Foundation.

Bonous-Hammarth, M. (2000). Pathways to success: Affirming opportunities for science, mathematics, and engineering majors. *Journal of Negro Education, 69*(1–2), 92–111.

Brown, M. C. (2003). Emics and etics of researching Black colleges: Applying facts and avoiding fallacies. In M. C. Brown & J. E. Lane (Eds.), *Studying diverse institutions: Contexts, challenges, and considerations* (pp. 27-40). New Directions for Institutional Research, no. 118. San Francisco, CA: Jossey-Bass.

Chang, M. J., Cerna, O., Han, J., & Saenz, V. (2008). The contradictory roles of institutional status in retaining underrepresented minorities in biomedical and behavioral science majors. *Review of Higher Education, 31*(4), 433–464.

Chen, X., & Weko, T. (2009). *Students who study science, technology, engineering, and mathematics (STEM) in postsecondary education* (NCES 2009-161). Washington, DC: Institute of Education Sciences, National Center for Education Statistics, U.S. Department of Education.

Cohen, A., & Brawer, F. (2008). *The American community college.* San Francisco, CA: Jossey-Bass.

Contreras, F., Malcom, L., & Bensimon, E. (2008). Hispanic serving institutions: Closeted identity and the production of equitable outcomes for Latino/a students. In M. Gasman, B. Baez, & C. S. Turner (Eds.), *Understanding minority-serving institutions* (pp. 71–90). Albany, NY: SUNY Press.

Dowd, A. (2003). From access to outcome equity: Revitalizing the democratic mission of the community college. *The ANNALS of the American Academy of Political and Social Science, 586*(1), 92–119.

Dowd, A., & Coury, T. (2006). The effect of loans on the persistence and attainment of community college students. *Research in Higher Education, 47,* 33–62.

Expanding Underrepresented Minority Participation. (2011). *Expanding underrepresented minority participation: America's science and technology talent at the crossroads.* Retrieved from http://www.nap.edu/catalog.php?record_id=12984#top

Fenske, R. H., Porter, J. D., & DuBrock, C. P. (2000). Tracking financial aid and persistence of women, minority, and needy students in science, engineering, and mathematics. *Research in Higher Education, 41*(1), 67–94.

Fries-Britt, S., Younger, T., & Hall, W. (2010a). How perceptions of race and campus racial climate impact underrepresented minorities in physics. In T. E. Dancy II (Ed.), *Managing diversity: (Re)visioning equity on college campuses* (pp. 181–198). New York, NY: Peter Lang.

Fries-Britt, S., Younger, T., & Hall, W. (2010b). Lessons from high achieving minorities in physics. In S. R. Harper, C. Newman, & S. Gary (Eds.), *Students of color in STEM: Constructing a new research agenda* (pp. 75–83). San Francisco, CA: Jossey-Bass.

Gasman, M., Baez, B., & Turner, C. S. (Eds.). (2008). *Understanding minority-serving institutions.* Albany, NY: SUNY Press.

Gasman, M., Lundy-Wagner, V., Ransom, T., & Bowman, N. (2010). Unearthing promise and potential: Our nation's historically Black colleges and universities [Special report]. *ASHE Higher Education Report: 35*(5).

Gloria, A. M., Castellanos, J., Lopez, A. G., & Rosales, R. (2005). An examination of the academic nonpersistence decisions of Latino undergraduates. *Hispanic Journal of Behavioral Sciences, 27*(2), 202–223.

Goodchild, F. (2004). The pipeline: Still leaking. *American Scientist, 92*(2), 112-113.

Hagedorn, L. S., Maxwell, W., & Hampton, P. (2002). Correlates of retention for African American males in community colleges. *College Student Retention, Research, Theory and Practice, 3,* 243–263.

Higher Education Act of 1965, retrieved from http://ftp.resource.org/gao.gov/89-329/00004C57.pdf

Hoachlander, E. G., Sikora, A. C., Horn, L., & Carroll, C. D. (2003). *Community college students goals, academic preparation, and outcomes* (NCES 2003-164). Retrieved from http://purl.access.gpo.gov/GPO/LPS126081

Hubbard, S., & Stage, F. K. (2009). Attitudes, perceptions, and preferences of faculty at Hispanic serving and predominantly Black institutions. *Journal of Higher Education, 80*(3), 270–289.

Hurtado, S., Alvarez, C., Guillermo-Wann, C., Cuellar, M., & Arellano, L. (2012). A model for diverse learning environments: The scholarship on creating and assessing conditions for student success. In J. Smart & M. Paulsen (Eds.), *Higher education: Handbook of theory and research* (pp. 41–122). New York, NY: Springer.

Hurtado, S., Eagan, M. E., Cabrera, N. L., Lin, M. H, Park, J., & Lopez, M. (2007). Training future scientists: predicting first-year minority student participation in health science research. *Research in Higher Education, 49*(2), 126–152.

Integrated Postsecondary Education Data System. (2010). U.S. Department of Education, National Center for Education Statistics. Retrieved from http://nces.ed.gov/ipeds/

Kim, M. K., & Conrad, C. (2006). The impact of historically Black colleges and universities on the academic success of African-American students. *Research in Higher Education, 47*(4), 399–427.

King, J. E. (2002). *Crucial choices: How students' financial decisions affect their academic success*. Washington, DC: American Council on Education.

Kinzie, J. (2002). Searching for a female Einstein: First year women negotiating introductory college chemistry (unpublished doctoral dissertation). Indiana University, Bloomington.

Lent, R. W., Brown, S. D., Sheu, H., Schmidt, J., Brenner, B. R., Gloster, C. S., ... Lyons, H. (2005). Social cognitive predictors of academic interests and goals in engineering: Utility for women and students at historically Black universities. *Journal of Counseling Psychology, 52*(1), 84–92.

Leslie, L. L., McClure, G. T., & Oaxaca, R. L. (1998). Women and minorities in science and engineering: A life sequence analysis. *Journal of Higher Education, 69*(3), 239–276.

Lopatto, D. (2004). Survey of undergraduate research experiences (SURE): First findings. *Cell Biology Education, 3*, 270–277

Li, X. (2007). *Characteristics of minority-serving institutions and minority undergraduates enrolled in these institutions* (NCES 2008-156). Washington, DC: National Center for Education Statistics, Institute of Education Sciences, U.S. Department of Education.

Maple, S. A., & Stage, F. K. (1991). Influences on the choice of math/science majors by gender and ethnicity. *American Educational Research Journal, 28*(1), 37–60.

Maton, K. I., & Hrabrowski, F. A. (2004). Increasing the number of African American PhDs in the sciences and engineering. *American Psychologist, 59*(6), 547–556.

May, G. S., & Chubin, D. E. (2003). A retrospective on undergraduate engineering success for underrepresented minority students. *Journal of Engineering Education, 92*(1), 27–39.

Mooney, G. M., & Foley, D J. (2011). *Community Colleges: Playing an Important Role in the Education of Science, Engineering, and Health Graduates*. NCES Info Brief. Retrieved from http://www.nsf.gov/statistics/infbrief/nsf11317/nsf11317.pdf

Morales, D. H. (2011). Transferring to a four-year Hispanic-serving institution: Latino student baccalaureate attainment (upublished doctoral dissertation). New York University, NY.

Museus, S. D., Jayakumar, U. M., & Robinson, T. (2011). Modeling racial differences in the effects of racial representation on 2-year college student success. *Journal of College Student Retention: Research, Theory and Practice, 13*(4), 549–572.

Museus, S. D., Palmer, R. T., Davis, R. J., & Maramba, D. C. (2011, March). Racial and ethnic minority students' success in STEM education [Special report]. *ASHE-Higher Education Report, 36*(6).

National Academy of the Sciences (NAS). (2010). *Expanding underrepresented minority participation: America's science and technology talent at the crossroads*. Washington, DC: National Academies Press.

National Center for Education Statistics. (2011). *Integrated postsecondary education data system*. Washington, DC: U.S. Department of Education, Institute of Education Sciences.

National Science Board. (2004). *Science and engineering indicators 2004*. Washington, DC: Author.

Oakes, J. (1990). Opportunities, achievement, and choice: Women and minority students in science and mathematics. *Review of Research in Education, 16*(2), 153–166.

National Science Foundation. (2009). *Women, minorities, and persons with disabilities in science and engineering: 2009.* Arlington, VA. Retrieved from http://www.nsf.gov/statistics/wmpd/

Palmer, R. T. (2010). The perceived elimination of affirmative action and the strengthening of historically Black colleges and universities. *Journal of Black Studies, 40,* 762–776.

Palmer, R. T., Davis, R. J., & Hilton, A. A. (2009). Exploring challenges that threaten to impede the academic success of academically underprepared African American male collegians at an HBCU. *Journal of College Student Development, 50,* 429–445.

Palmer, R. T., Davis, R. J., & Thompson, T. (2010). Theory meets practice: HBCU initiatives that promote academic success among African Americans in STEM. *Journal of college student development, 51*(4), 440–443.

Palmer, R. T., & Gasman, M. (2008). 'It takes a village to raise a child': The role of social capital in promoting academic success for Black men at a Black college. *Journal of College Student Development, 49*(1), 52–70.

Paulsen, M. B., & St. John, E. P. (2002). Social class and college costs: Examining the financial nexus between college choice and persistence. *Journal of Higher Education, 73,* 189–236.

Perna, L. W., Lundy-Wagner, V. C., Drezner, N. D., Gasman, M., Yoon, S., Bose, E., & Gary, S. (2009). The contribution of HBCUs to the preparation of African American women for STEM careers: A case study. *Research in Higher Education, 50*(1), 1–23.

Roksa, J., & Calcagno, C. (2008, June). *Making the transition to four-year institutions: Academic preparation and transfer.* CCRC Working Paper No. 13. New York, NY: Community College Research Center, Teachers College, Columbia University.

Shannon, H., & Smith, R. (2006). A case for the community college's open access mission. *New Directions for Community Colleges, 136* (Winter), 15–21.

Solórzano, D. G. (1995). The doctorate production and baccalaureate origins of African Americans in the sciences and engineering. *Journal of Negro Education, 64*(1), 15–32.

Stage, F. K. (2007). *Answering critical questions using quantitative data.* In F. Stage (Ed.), *Using quantitative data to answer critical questions* (pp. 5–16). San Francisco, CA: Jossey-Bass.

Stage, F. K., & Hubbard, S. M. (2009). Undergraduate institutions that foster women and minority scientists. *Journal of Women and Minorities in Science and Engineering, 15,* 77–91.

Stage, F. K., John, G., & Hubbard, S. M. (2011). Undergraduate institutions that foster Black scientists. In W. F. Tate & H. T. Frierson (Eds.), *Beyond stock stories and folktales: African Americans paths to STEM fields* (pp. 3–21). Bingley, England: Emerald Group Publishing.

Stage, F. K., Lundy-Wagner, V., & John, G. (2012). Minority serving institutions and STEM: Charting the landscape. In R. Palmer, D. Maramba, & M. Gasman (Eds.), *Fostering success of ethnic and racial minorities in STEM: The role of minority serving institutions* (pp. 16–32). New York, NY: Routledge.

Suitts, S. (2003). Fueling education reform: Historically Black colleges are meeting a national science imperative. *Cell Biology Education, 2,* 205–206.

Tinto, V. (1975) Dropout from higher education; A theoretical synthesis of recent research. *Review of Educational Research, 45,* 89–125.

Tsui, L. (2007). Effective strategies to increase diversity in STEM fields: A review of the research literature. *The Journal of Negro Education,* retrieved from FindArticles.com at http://findarticles.com/p/articles/mi_qa3626/is_200710/ai_n25139929/

U.S. Department of Education, National Center for Education Statistics. (2009). *Digest of education statistics, 2008* (NCES 2009-020). Washington, DC: Author.

Ware, N. C., & Lee, V. E. (1988). Sex differences in choice of college science majors. *American Educational Research Journal, 25*(4), 593–614.

Wolf-Wendel, L. E. (1998). Models of excellence: The baccalaureate origins of successful European American women, African American women and Latinas. *Journal of Higher Education, 69*(2), 144–172.

10
CREATING SUCCESSFUL PATHWAYS FOR ASIAN AMERICANS AND PACIFIC ISLANDER COMMUNITY COLLEGE STUDENTS (AAPIS) IN STEM

Dina C. Maramba

Supporting the successful participation of students of color in STEM is an imperative goal for institutions of higher education. This is important not only because of increasing racial and ethnic diversity among students in higher education, but also because of moral and ethical obligations to address the systemic inequities in STEM (Museus, Palmer, Davis, & Maramba, 2011). Colleges and universities have a responsibility to foster a learning environment that is accessible, supportive, and equitable for all students. Higher education institutions also have a valuable stake in the success of all students who plan to pursue STEM because of the eventual benefits and impact on the STEM workforce. For example, the Committee on Prospering in the Global Economy of the 21st Century (2007) emphasized that graduates who have the ability to be successful in the STEM workforce will contribute to a more productive national economy. In addition, student success in STEM increases the ability of the United States to compete in the global economy (Museus et al., 2011). With these concerns in mind, perhaps an issue in need of further attention is understanding how specific underrepresented racial and ethnic minorities (URMs) fare in STEM majors. However, the research in these areas is very limited especially with regard to examining experiences of Asian American and Pacific Islander (AAPIs) community college students in STEM related majors. Therefore, this chapter will focus on the role of community colleges in creating STEM pathways for AAPIs.

Indeed, there is truly a concern for the participation of students of color in STEM at all levels of education. Interestingly, while statistics have provided information regarding the lack of participation of African Americans, Latinos, and Native Americans in STEM (Museus et al., 2011), reports indicate that the participation of Asian American and Pacific Islander (AAPI) students in STEM is not one of concern. Specifically, data from the 2012 Pew report asserts that

AAPIs are the best educated of all groups in comparison to all U.S. adults. Given this information, the prevailing conclusion is that AAPIs need little attention in education (in general) and in STEM discourse (in particular). For example, aggregated data show that the proportion of science and engineering bachelor degrees awarded to AAPIs is higher than the AAPI population itself (9.3% to 4.6%, respectively). As such the common assumption is that support services or recruitment of this population in STEM is unnecessary (Museus et al., 2011).

Statistics such as these can be misleading; therefore it is necessary to take a closer look at the AAPI category itself. There are a number of issues to consider. First, it is important to acknowledge that the AAPI grouping is comprised of very distinct ethnic populations (National Commission, 2011; Reeves & Bennett, 2004). Thus, it is important to disaggregate data by AAPI subgroups in order to gain a more detailed understanding of the differential enrollment and success rates among this population. Moreover, when AAPI students are placed in the context of community colleges and STEM education, a different picture emerges compared to their counterparts in four-year institutions. Consequently, the goal of this chapter is to understand how AAPI community college students fit in the discourse of URMs and STEM. This chapter will cover five sections: (1) provide a brief description of the complexity of the AAPI category; (2) contextualize AAPIs in community colleges; (3) understand AAPIs' participation in STEM; (4) describe current initiatives and approaches that help address concerns surrounding AAPI community college students and STEM; and (5) provide implications for researchers, practitioners, and policy makers.

AAPI Ethnic Groups

According to the 2010 Census, AAPIs consist of 48 diverse ethnic backgrounds. The Asian American populations include but are not limited to the following: Asian Indian, Bangladeshi, Cambodian, Chinese, Filipino, Hmong, Japanese, Korean, Laotian, Pakistani, Taiwanese, Thai, and Vietnamese. The Pacific Islander populations include but are not limited to the following: Chamorro, Fijian, Guamanian, Mariana Islander, Native Hawai'ian, Saipanese, Samoan, Tahitian, Tongan, Micronesian, Melanesian (National Commission, 2008). The data on AAPIs is bimodal (Chew-Ogi & Ogi, 2002) that is, because of the vast diversity, the distribution of this population's characteristics tends to be on opposite extremes. Among many differences, they vary not only by ethnicity but language, immigration, generational status, and socioeconomic class. For example, with regard to language, 79% of Asian American students speak a language other than English at home; 43% for Pacific Islanders (National Commission, 2008). The language differences have serious implications for educational institutions as they need to be more cognizant of the unique needs of English language learners and provide support that values bilingualism and biculturalism.

As for immigration, it is important to understand that AAPI populations have varied immigration histories. For instance, due to the immigration Act of 1990, the United States allowed certain groups to enter the country through particular preferred qualification such as skilled professionals that serve the needs of the country (National Commission, 2008). Some AAPI ethnic groups entered the United States through refugee and asylee status. In addition, there are AAPI ethnic groups such as Native Hawaiians and groups from the U.S. territories (e.g., Samoa, Guam) who were already physically part of the United States. On a socioeconomic level, the disaggregated data show that a number AAPI groups are below the national poverty rate of 12.4% (National Commission, 2008). For instance, compared to the poverty rates of their AAPI counterparts (e.g., Japanese, 9.7%; Asian Indian, 9.8%), Southeast Asians, which include Hmong, Cambodian, Laotian, and Vietnamese have much higher poverty rates (37.8%, 29.3% 18.5%, and 16.6%, respectively). Moreover, Pacific Islanders, which include Marshallese, Samoan, Tongan, Native Hawaiian, and Guamanian have poverty rates as follows 38.3%, 20.2%, 19.5%, 15.6%, and 13.6% (National Commission, 2008).

AAPI Participation in Higher Education with Attention to Community Colleges

The dearth of research on AAPIs in higher education continues to be a concern. The lack of research has resulted in a population that is, for the most part, misunderstood and misrepresented (Chang, 2009). One of the stereotypical notions of AAPIs is that they are high achieving and successful in education. However, AAPI groups are at opposite ends of the spectrum with regard to educational achievement (Chew-Ogi & Ogi, 2002). Researchers agree that there are multiple reasons that attribute to this lack of understanding of the AAPI population. First, the model minority stereotype associated with AAPIs further reinforces misperceptions such as that *all* AAPIs are academically successful. Unfortunately, this misrepresentation has grave consequences regarding how educational institutions provide or not provide services for AAPI students (Maramba, 2008; Suzuki, 2002). Second, as discussed earlier, the fact that the AAPI category is an extremely diverse group has led to overgeneralization. The continued aggregation of such a large group deters educators from truly understanding the intricacies and differences of ethnicities within the AAPI racial group (Hune, 2002). That is, instead of treating AAPIs as a monolithic group, their immigration/generational differences, cultural differences, socioeconomic differences should be taken into account. Considering these variations will allow for a more accurate picture of the AAPI populations, research should emphasize the importance of critically disaggregating this category in order to better serve AAPI students.

For example, with regard to educational attainment of AAPI college attendees, more than four out of five Chinese, Japanese, Koreans, Asian Indian, and Pakistani earned at least a bachelor's degree (National Commission, 2011). In contrast, Southeast Asians who attended college but ultimately did not earn a degree are highly disproportionate (42.9% Cambodian, 47.5% Hmong, 46.5% Laotian, and 33.7% Vietnamese) compared to their AAPI counterparts (e.g., 8.2%, Asian Indian, 12.5% Chinese, and 12.7% Pakistani). Moreover, the educational attainment for the Pacific Islander population shows comparably higher rates of attending college but not earning a degree at 47.0%, 50.0%, 54.0%, and 58.1%, respectively (National Commission, 2011). Upon examining statistics further, larger educational disparities among AAPI subpopulation are revealed regarding pursuing any form of postsecondary education. Compared to their AAPI counterparts (25 years or older) who have not attended college (e.g., Asian Indians 20.4%; Japanese, 27.8%; Korean, 29.3%), larger proportions of Pacific Islanders and Southeast Asian Americans (25 years or older) have not enrolled or even completed any postsecondary education. For example, in 2006–2008, among Pacific Islanders (25 years or older), 57.9% Tongan, 56.8% Samoan, 53.0% Guamanian, and 49.3 % Native Hawaiian have not attended college. Along the same lines, among Southeast Asians (25 years or older), 65.8% Cambodian, 65.5% Laotian, 63.2% Hmong, and 51.1% Vietnamese have not attended college (National Commission, 2011).

AAPIs in Community Colleges

AAPIs are often misunderstood in higher education. For example, there is a predominant misperception that AAPIs are "overrepresented" in U.S. colleges and universities. This notion of overrepresentation is inaccurate for a few reasons. First, AAPIs account for only 6% of the U.S. college-going population. Second, two-thirds of the AAPI college-going population are heavily concentrated in only 200 institutions across the United States and attended college in only eight states in the year 2000. Moreover, approximately half of AAPIs attended college in California, New York, and Texas (National Commission, 2008). However, when we look specifically at the types of higher education institutions that AAPIs are attending, namely community college, the data paint a different picture.

Upon closer examination, the largest sectors of AAPI college students are concentrated in community colleges. Enrollment of AAPIs in public two-year colleges has been increasing at a faster rate than those enrolling at four-year institutions (National Commission, 2010). In particular, AAPI subgroups such as Laotians, Cambodians, Hmong, and Vietnamese are more likely to enroll in community college (Teranishi, 2010). In 1995, statistics show that AAPIs have consistently been heavily represented in community colleges compared to AAPIs in four-year institutions (44.6% and 40.1%, respectively). In 2005, of the total AAPI college-going population, 47.3% of AAPI students attended a public two-year college compared to 38.4% of AAPIs enrolled public, four-year institutions. AAPI community college students are also demographically different from their AAPI counterparts attending four-year institutions. The majority of AAPI community college students tend to be older (average age of 27.3 years), 62.9% are part time, and 31.7% have a delayed matriculation rate of two years or more (National Commission, 2010). In addition, 80% of AAPI community college students are found across only eight states and tend to be clustered in institutions located in California, Hawai'i, and New York (National Commission, 2010).

Furthermore, enrollment of AAPIs in public two-year colleges has the highest number of risk factors contributing to low persistence and completion as compared to AAPIs at four-year institutions (National Commission, 2010). For example, 5.1% of AAPI community college students have four or more risk factors compared to 2.8% of AAPIs at four-year institutions. These risk factors include: "delayed enrollment, lack of a high school diploma (including GED recipients), part-time enrollment, having dependents other than spouse, single parent status, and working full-time while enrolled (35 hours or more)" (National Commission, 2010, p. 16). In addition, although the cost to attend community colleges is relatively low compared to other educational institutions, financing their education is an ongoing concern for AAPIs in community colleges. As many tend to be low income and first generation college, AAPI community college students, compared to other racial and ethnic groups, have a higher likelihood of having difficulties financing their education. One reason for this difficulty is although AAPI community college students have high financial need (45.5% are more than $2,000 in need and 10.6% are more than $8,900 in need), they are less likely to apply for federal financial aid compared to other racial or ethnic groups (National Commission, 2010).

Another important area of concern for AAPIs in community colleges is the lack of extant literature and research on this particular population. Although there is already a dearth of published empirical research on AAPIs in four-year institutions, there is even less empirical research on AAPIs who attend community colleges. Similar to the research on AAPIs in four-year institutions, studies on AAPIs in community colleges have a similar call for the need to understand the diverse AAPI ethnicities within the category. For example, based on her bibliography of 12 references, consisting of reports, case studies, and literature reviews on AAPIs in community college, Liu (2007) further reasoned that by addressing disaggregation of the AAPI population, and the model minority stereotype, research will be more useful and provide a better understanding of AAPIs in community colleges. In their review of literature on AAPIs in community colleges, Lew, Chang, and Wang (2005) also stressed the scarcity of research in this area. Based on their review, they concluded that APA community college students are diverse and must not be looked upon as a homogeneous group. They also explained that the existing literature on AAPIs in four-year institutions helped serve as a guide to address the gaps in the literature about AAPI community college students.

Orsuwan (2011) conducted a quantitative study in Hawaii on the community college learning experiences of specific AAPI ethnicities. He found that there were significant differences among academic integration, sense of belonging and family income for Japanese, Chinese, Filipino, and Hawaiian ethnic groups. More specifically, the academic integration of Japanese (from the upper social categories of Hawai'i) and Filipinos and Hawaiians (from the lower social categories of Hawai'i) showed statistical significance. Orsuwan attributed this finding to differences in socioeconomic status, education attainment, and arrival to Hawai'i amongst these groups. Similarly, Orsuwan and Cole (2007) disaggregated the AAPI community college student population and found that the interaction with the college experience made race and ethnicity "more dynamic." Among their findings, they explained that Filipino academic integration played a more important role in their educational satisfaction compared to the other AAPI groups in the study. In addition, for Hawaiian students (the second lowest income group in the study), household income was negatively associated with their educational satisfaction. Bottrell, Banning, Harbour, and Krahnke (2007) conducted a study on Vietnamese community college students in which they found that participants heavily used heuristic knowledge to help themselves navigate through their academic and social experiences and that they heavily attributed their Vietnamese community to assisting in their heuristic knowledge.

On an institutional level, Laanan and Starobin (2004) discussed the importance of the potential role of the federal designation of AAPI serving institutions. They further discussed how the AAPI serving designation could benefit AAPI students especially for those attending two-year colleges. Specifically, they emphasized that because of the characteristics of the environment and demography of two-year community colleges, these institutions are in a unique position to serve in the development of human and social capital for the fast growing AAPI population.

AAPI Participation in STEM

Although there is very little information on AAPIs in STEM at community colleges, we can speculate through the statistics and studies of AAPI achievement at four-year institutions that similar issues may be occurring at the community college level. Similar issues that both AAPI students at four- and two-year institutions may experience include financial aid issues, being a first generation college student, and being misperceived as overrepresented in the STEM majors. However, we can also conjecture that these issues may be magnified given the higher risk factors of AAPI community college students (National Commission, 2010).

As discussed earlier, when AAPIs are examined regarding STEM issues, they are often portrayed as being overrepresented or already successful in these fields. As a result, the STEM discourse often focuses on the comparison of AAPI and White populations to the low numbers of other URMs (Latinos, African Americans, and Native Americans). To counter the misrepresented notion all AAPI populations' success in STEM, a few forums have attempted to address the concern. For instance, a national two-day workshop sponsored by the Asian American and Pacific Islanders' Coordination Committee of the National Science Foundation was held in 2003. Comprised of professions and NSF staff, they discussed ways that NSF can broaden the participation of AAPIs in NSF programs and the STEM workforce (National Science Foundation, 2003). Although this workshop addressed these concerns, for the most part, AAPIs are often excluded in discussions regarding underrepresented populations in STEM. Existing statistics on specific AAPIs and STEM majors must once again be closely examined to create a better understanding of these issues.

Although there are a number of AAPIs who major in STEM, only 21.8% of degrees awarded to AAPI students are in STEM fields (National Commission, 2008). A larger proportion of AAPIs who received bachelor's degrees earned them in management (28.8%), which is closer to the national average of 33.7%. and in the social sciences (26.1%), which is higher than the national average of 19.5% (National Commission, 2008). Furthermore, 2006–2008 statistics that examine specific AAPI ethnicities with regard to occupation indicate that less than 10% of Native Hawaiian, Tongan, Gauamanian/Chamorro, Samoan, Vietnamese, Cambodian, Hmong, and Laotian are employed in health and STEM occupations (National Commission, 2010).

Aside from previously discussed topics as the model minority and the failure to acknowledge AAPIs as a heterogeneous group, another one of the major issues surrounding the dearth of AAPI research and STEM is the lack of inclusion and recognition of this group as an underrepresented racial and ethnic minority group in STEM. For example, the 2011 National Science Foundation report on women, minorities, and persons with disabilities in science and engineering clearly stated that

> Asians are not considered underrepresented because they are a larger percentage of science and engineering degree recipients and of employed scientists and engineers than they are of the population. Subgroups of Hispanics and Asians may vary in terms of under- or overrepresentation in science and engineering.
>
> (p. 2)

Although the report acknowledged that subgroups of Asians may be under or overrepresented, this discrepancy is not discussed further in the report.

Museus and colleagues (2011) explained that aggregated statistics such as math performance, SAT math scores, and STEM degree aspirants who completed a STEM degree, appear to illustrate that AAPIs consistently do well in these areas. However, these numbers are misleading because they are seen in the aggregate. When we examine AAPIs who enter community colleges, the majority of them are underprepared in basic academic areas. In 2003, more than half (55.2%) of AAPI community college students were not prepared in math, most of whom did not complete an Algebra II course in high school. In addition, one in five AAPI students needed to take remediation classes in English (National Commission, 2011).

Qualitative studies are also helpful in contextualizing specific AAPI ethnic groups' experience in STEM majors. For example, Palmer, Maramba, and Dancy (2011) conducted a study on Southeast Asian American (SEAA) and Black student STEM majors at a four-year research university and found that this population highly benefitted from support services, tutoring, academic advising, peer support, and involvement in STEM related activities. These activities also served as encouragement to pursue STEM careers. These findings are important as they urge higher education institutions to foster more supportive and less intimidating environments for students of color in order to successfully graduate as STEM majors. As noted earlier, researchers urge a more critical view by understanding the varying levels of achievement of the AAPI subgroups, including whether they are Asian international and Asian American students (National Commission, 2010). A closer examination of these numbers reveals a very different and alarming picture with regard to AAPI educational achievement and their completion in STEM majors. The following section describes examples of current initiatives that help address AAPI community college students in STEM.

Current Initiatives and Efforts that Address AAPI Community College Students in STEM

This section briefly describes two selected current initiatives that help address the issues faced by AAPIs in STEM: (1) Asian American and Native American Pacific Islander Serving Institutions (AANAPISIs) and (2) the Asian Pacific Islander American Scholarship Fund (APIASF, n.d.). Although there are other efforts that help address these issues, the following serve only as examples of ways institutions and initiatives focus on creating more supportive educational environments and producing positive outcomes for AAPI populations in all majors, including STEM disciplines.

Asian American and Native American Pacific Islander Serving Institutions

Asian American and Native American Pacific Islander Serving Institutions (AANAPISs) are federally funded institutions designated as a Minority Serving

Institution (MSI) that aim to increase the access and positive educational outcomes for low income and underserved AAPI students. AANAPISIs, the most recently established MSI, is a relatively new initiative that was included as part of the College Cost Reduction and Access Act of 2007 (National Commission, 2010; Park & Teranishi, 2008; Teranishi, Maramba, & Ta, 2013). For an institution to be considered an AANAPISI, it must apply for the federal designation. A designated AANAPISI must have at least 10% enrollment of AAPI students, a minimum threshold of low income students, and lower educational and general expenditures per student. Additionally, to receive funding, a designated AANAPISI must also compete for a grant that requires applicants to create a plan that addresses the needs of AAPIs at their particular campus. Although each funded AANAPISI may differ in its approach to addressing AAPI concerns, all grantees that have been funded focus on academic and support services, leadership and mentorship opportunities, and research and resource development. As of 2011, there are approximately 116 institutions in the country that are eligible to be an AANAPISI. However, at the present time, there are only 52 institutions that have applied and are approved as a designated AANAPISI. Thus far, there are a total of 15 institutions that have been funded to receive aid from the federal government. These higher education institutions are located in California, Hawaii, Illinois, New York, Massachusetts, Maryland, Texas, Washington, and the U.S. territory of Guam. It is also important to note that more than half of these institutions are public two-year colleges (Teranishi et al., 2013). The fact that the majority of the funded AANAPISIs are two-year colleges is critical. Given the AAPI educational disparities and concerns at the community college level, the designation and funding are essential in providing support and addressing the unique needs of particular AAPI ethnicities.

A helpful example of how AANAPISIs have served as a mutable site of opportunity for encouraging AAPI STEM participation is the effort put forth by the City College of San Francisco (CCSF), one of the institutions included in the first cohort of AANAPISIs to be funded from 2008–2011 (Teranishi et al., 2013). Although not all funded AANAPISIs focus on STEM, CCSF is one such institution that determined that this was as an area of concern for their campus. As one of the oldest and largest community colleges in the nation, CCSF enrolls approximately 100,000 students with over 4,700 courses throughout nine campuses. The demographics of CCSF reflect a diverse population of students. The AAPI ethnicity breakdown is as follows: Asian Americans 27%, Filipino 7%, Pacific Islander 1%, and Southeast Asian 3%. Additional populations include White (27%), Latinos (15%), African American (8%), American Indian/Alaskan Native (1%), Other (3%), and Unknown/No Response (9%). Eighty-three percent are part-time students and 30% receive financial aid. With regard to language, 23% of the 54,000 degree-seeking students are Asian Americans and Latinos who are non-native English speakers. While 74% of the overall

CCSF population begins in developmental education, they are at higher risk of not completing a degree.

CCSF recognized that a number of AAPI students struggle in the STEM fields. Specifically, they found that compared to other Asian American subgroups and the White population, Filipino and Pacific Islander students fared much lower with regard to degree attainment and transfer rates in STEM. Upon further examination of data relevant to STEM disciplines, CCSF found only 56% of AAPI students who took remedial math continued to the next level and only 30%-35% complete math courses transferable to the university level. The passing rate for college chemistry for AAPIs was 46%-47%.

In acknowledging these disparities among AAPI students, CCSF addressed these issues by creating a STEM Achievement Project (ASAP). This project aimed to increase the transfer rates for disadvantaged students which also included AAPI students with particular attention to Southeast Asians, Filipino Americans, and Pacific Islanders. CCSF created an ASAP center that provided space for a computer lab, study area, and office space for instructional and counseling faculty to provide services for the students. Several programs were also created which included recruitment and outreach activities to target specific populations for the ASAP center. Participating students benefitted from a book loan program, additional academic and instructional services, and opportunities to attend various STEM related conferences in the local area.

While CCSF is still in the process of analyzing the 3 years of data in determining the success of the program, there are indeed a number of positive outcomes thus far. In spring 2012, 42% of the STEM participants successfully transferred to a four-year institution. Moreover, compared to nonparticipants in the ASAP center who had completion rates of 15%-20%, AAPI students who started as STEM majors and participated in the ASAP program services, had completion rates between 50%-60% (Teranishi et al., 2013).

Asian Pacific Islander American Scholarship Fund National Commission on Asian American and Pacific Islander Research in Education

The Asian Pacific Islander American Scholarship Fund (APIASF) fund was established to bring together various resources that according to Neil Horikoshi, the APIASF President and Executive Director, "creates opportunities for deserving Asian American and Pacific Islander students to access, complete and succeed after post-secondary education" (Asia Pacific Islander Scholarship Fund, para. 1, n.d.). APIASF, the largest nonprofit organization of its kind in the country, assists AAPI students pursuing postsecondary education and has allocated more than $60 million in scholarships since 2003. Among their successful outcomes, APIASF has played an important role in providing financial assistance for AAPI community college students. Since 2005, APIASF scholarships have included recipients from all 48 AAPI ethnicities and from all of the states including the U.S. territories. In addition, 58% of the 2011–2012 scholarship awardees came from families living at or below the poverty line. This organization has helped counter the notion that all AAPIs are not in need of financial aid or academic support. APIASF also builds partnerships with funding sources throughout the country to provide the yearly scholarships for AAPI students.

Another important aspect of APIASF is its collaboration with the National Commission on Asian American and Pacific Islander Research in Education (CARE). This partnership is critical as it connects research to policy and practice to help inform not only the general public but also higher education institutions and educational policy makers at the national level. Since 2008, CARE has produced three major research reports that are critical in addressing the postsecondary concerns and challenges faced by AAPI students (see Asian Pacific Islander Scholarship Fund, n.d.). Moreover, the CARE reports have played a pivotal role in presenting information that is used for policy makers. It is important to note that these reports have helped model the significance of disaggregating the AAPI groups to reveal a clearer picture of the educational disparities among them. For example, the research and statistics in the report have helped bring to light the disparities of the AAPI ethnic populations at the community college level with regard to STEM related concerns and AAPI students.

Implications for Researchers, Practitioners, and Policy Makers

The following section is a discussion on how the previously presented information on AAPI community college students and their participation in STEM has implications for policy makers, researchers, and practitioners.

First, and probably one of the most critical aspects of having a clearer understanding of AAPI community college students and STEM is to recognize that the AAPI category is made up of diverse ethnic groups. Among a number of differences, these populations have very distinct immigration histories, languages, socioeconomic class, and educational attainment. As maintained by scholars and researchers, disaggregating the data is a primary concern. This, however, cannot be done without a national effort to collect data on AAPIs in a more critical and carefully planned way. Currently, the census has recognized 48 AAPI ethnic groups. However, the collection of data within and across educational institutions is not consistent. Some institutions collect and disaggregate AAPI data but the concern is that this process does not take place at all colleges and universities. A current effort at the national level is a request for information (RFI) by the Department of Education. The intention of the RFI is to collect data that pertains to "promising practices and policies regarding existing education data systems and models that disaggregate data on subgroups within the ANHPI [Asian and Native Hawaiian and Other Pacific Islander] student population" (Department of Education, Federal Register, 2012). In other words, the Department of Education would like to have a better understanding of how various institutions across the country collect data on racial and ethnic subcategories of AAPI students. This is important because it is one of the first attempts at the federal level to acknowledge that disaggregating data within the AAPI population is critical. The outcome of this effort is yet to be known as the deadline for State Educational agencies (SEAs), local educational agencies (LEAs), schools, and institutions of higher education (IHEs), to submit their best practices or approaches is in the summer of 2012. Attempts such as this one must continue at the national policy making level in order for more consistent and effective collection of data on AAPI ethnic groups.

Because of the lack of overall acknowledgment by most institutions that AAPIs are underserved and underrepresented in STEM, national initiatives such as this RFI are essential as they may have a positive influence in the way national foundations and grants view AAPIs. For example, as noted earlier, the National Science Foundation explained that Asians are not considered underrepresented because there are a larger percentage of those in engineering and the sciences than the population itself. It may be safe to conclude that such assumptions of overrepresentation in STEM are based on aggregated AAPI data. Thus, assumptions such as these often result in the exclusion of AAPIs from receiving grants, fellowships, or other forms of aid. Moreover, these assumptions also disregard AAPI ethnic groups that have lower and varying levels of academic achievement.

In terms of research, studies that involve examining the aggregate data and the disaggregated data on AAPI community colleges and STEM must continue. In other words, while the aggregate data may be useful in some instances, studying specific AAPI ethnic groups such as Southeast Asian Americans, Filipino Americans, Native Hawaiians, and Pacific Islanders brings a clearer understanding of disparities among AAPIs. In addition, effective utilization of varied research methodologies is important. While quantitative and statistical data in research studies is important, qualitative empirical data is equally important (Maramba, 2011). Although quantitative data may explain trends with certain AAPI ethnic groups in STEM, qualitative data illuminates why these trends are occurring. This is particularly important in order to understand AAPI ethnic populations in STEM that are underserved and underrepresented in community colleges. For example, the qualitative study conducted by Palmer et al. (2011) provided more context to the experiences of Southeast Asian Americans student in STEM disciplines. Moreover, longitudinal and comparative research studies that document inequities across STEM AAPI ethnic groups provide additional pertinent information.

There are also a number of implications for student affairs practice. At the postsecondary level, more specifically at the community college level, student affairs practitioners have much potential to influence positive outcomes for AAPIs in STEM. There are various ways that this can take place especially with regard to facilitating an environment that considers culture, socioeconomic class, academic support, and eventually successful transfer to a four-year institution. First, community college student affairs practitioners can begin or continue their efforts in recognizing that AAPIs are not a monolithic group. Another area they must consider is that their participation in STEM is in fact a concern because of the underrepresentation of particular AAPI ethnic groups such as Southeast Asian Americans, Filipino Americans, and Pacific Islanders. Keeping these points in mind will have a positive influence on helping student affairs staff to better understand the concerns and challenges that AAPI students face. In turn, student services will be more effective in providing academic advising, counseling, and financial aid among other support mechanisms within the university. STEM related services and activities that incorporate gaining "transfer student capital" (Laanan, Starobin, & Eggleston, 2010-2011) can also benefit AAPI student success by providing ways in which AAPI students can gain important information to be retained as STEM majors. Transfer student capital as defined by Laanan et al. (2010–2011) is "where cumulative knowledge and experiences of higher education environment promote successful adjustment when students transfer to a 4-year institution" (p. 180). An example of a model project is one instituted at Iowa State University (ISU) called the STEM Student Enrollment and Engagement through Connections (SEEC). In conjunction with the nearby community college, Des Moines Area Community College (DMACC) aims to increase the number of pre engineering students, specifically women and minorities, to transfer and eventually graduate from Iowa State University. Through "learning villages," enhanced academic advising, mentoring programs, and building a strong relationship between DMACC and ISU, the goal is to increase pre-engineering students' engagement with their major and the college environment. Therefore, community college student affairs practitioners need to begin or continue to be more intentional with encouraging, supporting, and helping increase AAPI students' potential to succeed in STEM disciplines by eventually transferring to a four-year institution (see STEM Student Enrollment, n.d.).

Another important issue to consider at the community college level is to address the area of articulation agreements with four-year institutions. There may even be a possibility of creating a pre-STEM major since most community colleges do not have such a major. This is another collaborative opportunity that can take place between community colleges and four-year institutions as it may be beneficial for students by providing support and increasing the possibility of continuing and eventually completing a STEM major.

Conclusion

It is clear that more research on STEM participation of racial and ethnic minorities needs to continue. In their comprehensive review of racial and ethnic minorities in STEM education, Museus et al. (2011) contend that although the majority of research on minority students in STEM concentrates on four-year institutions, the role of community colleges is critical in STEM education for racial and ethnic minorities. The current literature, however, on STEM students of color in community colleges is scant. Likewise, it is even more uncommon to find STEM literature that includes AAPIs as a population of concern for STEM disciplines in general. Despite the stereotypical notion that AAPIs are well represented in STEM fields, we must continue to examine closely statistics that are readily available at the current time.

The case of AAPI community colleges and their involvement with STEM disciplines is a critical issue that must continue to be explored. The complexity of the AAPI population in general is the vast diversity of its ethnic groups and the need for disaggregation in order to understand the intricacies of immigration history, socioeconomic class, and educational attainment among many important factors. The mere acknowledgment that the disaggregated data of the AAPI ethnic groups is imperative will encourage educational institutions as well as national and federal organizations that provide financial support to include AAPIs in the STEM discussion about the continued systemic disparities that continue to plague underrepresented and underserved communities. Incentives at the federal level may further encourage immediate action. As presented earlier, the efforts of AANAPISIs and nonprofit organizations such as APIASF and CARE have certainly brought these issues to light. Community colleges play an important role as they are institutions that serve the majority of AAPIs at the postsecondary level, and recognize the importance of increasing their participation and successful completion of STEM disciplines.

References

Asian Pacific Islander Scholarship Fund. (n.d.). Retrieved from www.apiasf.org/welcome.html

Bottrell, C. A., Banning, J. H., Harbour, C. P., & Krahnke, K. (2007). The role of heuristic knowledge in Vietnamese American students' success at a midwestern community college. *College Student Affairs Journal, 27*(1), 116–135.

Chang, M. (2009). Asian evasion: A recipe for flawed solutions. *Diverse Issuues in Higher Education, 25*(7), 26.

Chew-Ogi, C. & Ogi, A. Y. (2002). Epilogue. In M. K. McEwen, C. M., Kodama, A. N. Lee, S. Lee, & C. T. H. Liang (Eds.), *Working with Asian American college students* (New Directions for Services, No. 97, 91-96). San Francisco, CA: Jossey-Bass.

Committee on Prospering in the Global Economy of the 21st Century. (2007). *Rising above the gathering storm: Energizing and employing American for a brighter future.* Washington, DC: National Academies Press.

Department of Education. (2012). Federal Register, Vol 77, no 87, Friday, May 4, 2012/Notices Docket ID ED-2012-OESE-0009.

Hune, S. (2002). Demographics and diversity of Asian American college students. In M. K. McEwen, C. M., Kodama, A. N. Lee, S. Lee, & C. T. H. Liang (Eds.), *Working with Asian American college students* (New Directions for Services, No. 97, pp. 11–20). San Francisco, CA: Jossey-Bass.

Laanan, F. S., & Starobin, S. S. (2004). Defining Asian American and Pacific Islander-serving institutions. *New Directions for Community Colleges, 127,* 49–50.

Laanan, F. S., Starobin, S. S., & Eggleston, L. E. (2010–2011). Adjustment of community college students at a four-year university: Role and relevance of transfer student capital for student retention. *Journal of College Student Retention, 12*(2), 175–209.

Lew, J. W., Chang, J. C., & Wang, W. W. (2005). UCLA community college review: The overlooked minority: Asian Pacific American students at community colleges. *UCLA Community College Review, 33*(2), 64–84.

Liu, A. (2007). UCLA community college bibliography: Asian Americans in community colleges. *Community College Journal of Research and Practice, 31,* 607–614.

Maramba, D. C. (2008). Immigrant families and the college experience: Perspectives of Filipina Americans. *Journal of College Student Development, 49*(4), 336–350.

Maramba, D. C. (2011). The importance of critically disaggregating data: The case of Southeast Asian American college students. *AAPI (Asian American Pacific Islander) Nexus Journal: The Role of New Research Data, & Policies for Asian Americans, Native Hawaiians, & Pacific Islanders, 9*(1 & 2), 127–133.

Museus, S. D., Palmer, R. T., Davis, R. J., & Maramba, D. C. (2011). Racial and ethnic minority students' success in STEM education [Special issue]. *ASHE-Higher Education Report Series, 36*(6).

National Commission on Asian American and Pacific Islander Research in Education. (2008). *Asian Americans and Pacific Islanders facts, not fiction: Setting the record straight.* New York, NY: Author.

National Commission on Asian American and Pacific Islander Research in Education. (2010). *Federal higher education policy priorities and the Asian American and Pacific Islander community.* New York, NY: Author.

National Commission on Asian American and Pacific Islander Research in Education. (2011). *The relevance of Asian Americans & Pacific Islanders in the college completion agenda.* New York, NY: Author.

National Science Foundation. Asian American and Pacific Islander Coordinating Committee. (2003, November). *Asian Americans and Pacific Islanders' issues: The challenges of success.* Workshop proceedings. Arlington, VA.

National Science Foundation. (2011). *Women, minorities, and persons with disabilities in science and engineering: 2011.* Arlington, VA: Author.

Palmer, R. T., Maramba, D. C., & Dancy, T. E. (2011). A qualitative investigation of factors promoting the persistence and academic success of students of color in STEM. *Journal of Negro Education, 80*(4), 491–504.

Pew Research Center. (2012, June 19). *The rise of Asian Americans.* Retrieved from http://www.pewsocialtrends.org/2012/06/19/the-rise-of-asian-americans/1/

Orsuwan, M. (2011). Interaction between community college processes and Asian American and Pacific Islander subgroups. *Community College Journal of Research and Practice, 35*(10), 743–755.

Orsuwan, M., & Cole, D. (2007). The moderating effects of face/ethnicity on the experience of Asian American and Pacific Islander community college students. *Asian American Policy Review, 16,* 61–86.

Park, J. J., & Teranishi, R. T. (2008). Asian American and Pacific Islander serving institutions. In M. Gasman, B. Baez, & C. S. V. Turner (Eds.), *Understanding minority serving institutions* (pp. 111–126). Albany, NY: State University of New York Press.

Reeves, T. J., & Bennett (2004). *We the people: Asians in the United States.* Washington, DC: U.S. Census Bureau.

STEM Student Enrollment and Engagement through Connections. Iowa State University. (n.d.). Retrieved from http://www.eng.iastate.edu/seec/about.shtml

Suzuki, B. H. (2002). Revisiting the model minority stereotype: Implications of student affairs practice and higher education. In M. K. McEwen, C. M., Kodama, A. N. Lee, S. Lee, & C.

T. H. Liang (Eds.), *Working with Asian American college students* (New Directions for Services, No. 97, pp. 21–32). San Francisco, CA: Jossey-Bass.

Teranishi, R. T. (2010). *Asians in the ivory tower: Dilemmas of racial inequality in American higher education*. New York, NY: Teachers College Press.

Teranishi, R. T., Maramba, D. C., & Ta, M-H. (2013). Asian American Native American Pacific Islander serving-institutions (AANAPISIs): Mutable sites of intervention for STEM opportunities and outcomes. In R. T. Palmer, D. C. Maramba, & M. Gasman (Eds.), *Fostering success of ethnic and racial minorities in STEM: The role of minority serving institution* (pp. 168–180). New York, NY: Routledge.

11

CONSTRAINTS AND OPPORTUNITIES FOR PRACTITIONER AGENCY IN STEM PROGRAMS IN HISPANIC SERVING COMMUNITY COLLEGES

Megan M. Chase, Estela Mara Bensimon, Linda Taing Shieh, Tiffany Jones, and Alicia C. Dowd

Introduction

The underrepresentation of Latinos/as in majors leading to degrees and professions in engineering, mathematics, computer science, and the biological and physical sciences has been well documented (e.g., Chang, Cerna, Ham, & Sa'enz, 2008; Garcia & Hurtado, 2011; Hurtado, Cabrera, Lin, Arellano, & Espinosa, 2009). Because of the sheer concentration of Latinos/as in the two-year college sector, community colleges have a crucial role in increasing their representation among science, technology, engineering, and mathematics (STEM) undergraduate degree holders. In this chapter we consider the role of community colleges in advancing educational opportunity in STEM for Latinos/as from the perspective of institutional agents. Institutional agents are defined as individuals who hold relatively high-status positions in society or institutions, and are in the position to provide students with key forms of social and institutional support (Stanton-Salazar, 1997, 2001). Until recently, the research focus on Latinos/as student success generally, and in STEM more specifically, has been almost entirely on their aspirations, behaviors, and other demographic characteristics as predictors of retention, achievement, and degree attainment (Cabrera, Burkum, & La Nasa, 2005; Cabrera, La Nasa, & Burkum, 2001), rather than on the actions of institutional actors to promote student achievement.

A consequence of associating success in STEM with student characteristics, behaviors, aspirations, and prior academic experiences is that Latinos/as often lack at least one of the main predictors of likely success in STEM (i.e., intensity of high school science and math courses). Thus, one of the predominant strategies employed by colleges to compensate for Latinos/as past academic experiences is through grant-funded programs that provide students with extra academic support and involve them in activities to build their confidence and identity as future scientists (Tsui, 2007). Undeniably, without the existence of programs such as Mathematics Engineering Science Achievement (MESA), Ronald E. McNair Scholars, Student Preparation for Academic Research Careers (SPARC), Minority Access to Research Careers Undergraduate Student Training in Academic Research (MARC * USTAR), and Louis Stokes Alliance for Minority Participation (LSAMP), inequality in STEM would be even greater than it is already.

However, special programs that serve small groups of students are not sufficient to break the cycle of Latinos/as underrepresentation in STEM. The magnitude of Latinos/as underrepresentation and inequality in STEM requires greater and broader involvement by STEM faculty and administrative leaders. In the long run, developing institution-wide capacity to increase participation and success in STEM fields for Latinos/as is a better investment than relying on programs and services financed through private foundations or government agencies. Special programs that depend on external funds are vulnerable to changing priorities and downturns in the economy. Moreover, these programs often live in the periphery, outside the academic core, and will not transform the practices of the majority of faculty members with whom Latinos/as will come into contact. Thus, students who are not among the program participants may find themselves in classrooms that do not provide them with the same level of support and validation (Rendón, 1994) as they have experienced in special programs.

Additionally, research studies have established the importance of out-of-class experiences for STEM majors such as involvement in undergraduate research, participation in conferences, and internship experiences (Hurtado et al., 2009). However, to make these experiences available to more than just a few Latinos/as, colleges need to build a critical mass of faculty and administrators who are invested in the success of Latinos/as. Yet most studies in STEM tend to assume a *student agency* stance (i.e., What can students do to gain access to faculty or other important resources to be successful? What perceptions do students have of their interactions with faculty and other STEM resources?). Few studies have examined the role of *faculty, administrator* or *staff agency* in supporting Latino/a students within and outside of the classroom or the ways that they integrate students into their professional networks and provide institutional support and resources. Since its founding in 2000, the research and action agenda that drives the work of the Center for Urban Education has focused on the development of institutional agency to produce equity in educational outcomes for racial and ethnic groups with a history of underrepresentation in higher education. As part of this agenda, we have studied the enactment of agency by faculty, staff, and administrators in Hispanic Serving Institutions (HSIs) specifically in relation to Latinos/as in STEM. It is important to study agency within HSI environments, as these institutions are well positioned to increase the academic success of Latinos/as by virtue of the large numbers of Latinos/as they serve.

Drawing on data from a larger case study of two-year and four-year institutional pairs,[1] we describe what motivates faculty, administrative leaders, and staff to offer themselves and their resources to facilitate success in STEM among Latinos/as. More specifically, we examine individuals' awareness of inequities in opportunity and their motivation to invest personal effort and resources to create opportunities for Latinos/as to successfully major in STEM fields. This study builds on a previous policy brief in which we introduced the concept of institutional agents in four-year institutions (Bensimon & Dowd, 2012). Our purpose is to continue to build on the concept of institutional agents with a focus on community colleges and to illustrate how individuals—acting alone and collectively—can take a more active role in advancing Latinos/as participation in STEM.

The concept of "agency," in critical theory, connotes human action to transform social conditions that create or perpetuate injustice based on gender, race, sexual orientation, socio-economic status, or other characteristics that are markers of "otherness" (Freire, 1970; Seo & Creed, 2002). As such, the action taken challenges and/or dismantles institutionalized structural arrangements that reproduce inequality. We use the term *agency* more generally to distinguish individuals who act to channel resources and educational opportunities to those student communities which have been historically under-represented in STEM fields and have not benefitted from the support and resources available to wealthier students. In previous research (Bensimon, 2007; Bensimon & Dowd, 2009), we have used the term *institutional agent* (Stanton-Salazar, 1997, 2001, 2011) to distinguish faculty members who act on behalf of underrepresented students. Whether referring to these individuals as exercising agency or as institutional agents, it is important to underscore that these individuals' actions on behalf of Latinos/as may not be motivated by commitment to social justice, awareness of institutionalized inequality, or an understanding of the historical, economic, and political conditions that have contributed to the lower levels of educational attainment among Latinos/as, particularly in the U.S.-Mexico border states.

Methodology

To describe the ways in which faculty, administrative leaders, and staff in two-year HSIs exercise agency on behalf of Latino/a students in the STEM fields, we drew on data from a larger case study of two-year and four-year institutional pairs conducted in July 2008. These pairs were purposefully sampled through a series of steps intended to identify exemplary and rich cases (Patton, 1990) of individuals within institutions who enacted multiple roles of agency on behalf of Latino/a students. Case selection was based on a multiple regression analysis of the federal Integrated Postsecondary Education Data System (IPEDS) data and institutional web-based document analysis to identify universities offering relatively high levels of institutional support to Latino/a students. Institutional support for Latinos/as was defined as (a) practices or institutional policies that characterize a HSI culture associated with exemplary STEM bachelor degree production and (b) evidence that an institution interprets or enacts state transfer policies to promote transfer equity and bachelor's degree attainment for Latinos/as. Three, four-year universities and their primary community college feeder institutions were selected as case study sites. Data collection included site visits to each of the institutions and interviews with over 100 faculty, staff, administrators, and students; observations of program activities; and document analysis. This chapter is based on interviews with 12 STEM faculty members, administrators, and academic affairs staff from West Coast College (WCC) and Southwest Community College (SCC). Using the qualitative software Atlas.ti, we coded the transcripts based on the following themes: motivation to support Latino/a STEM students, types of supports provided to Latino/a STEM students, and individuals and institutional barriers to supporting Latino STEM students. The colleges are similar in size, West Coast College enrolls approximately 25,000 students, and Southwest Community College enrolls approximately 20,000 students. The racial and ethnic composition of their student body is also similar, and in both colleges Latinos/as are the largest population. Table 11.1 lists the 12 participants based on their position and Table 11.2 lists the participants by their fictitious name and provides their race/ethnicity, gender, and generic position.

Findings

The results are organized into two sections. The first section describes barriers to the exercise of agency on behalf of Latinos/as and the second section features actions that illustrate ways in which a handful of individuals exercised agency. Surprisingly, even though the focal colleges were identified as Hispanic Serving, the Hispanic Serving identity was not salient. For example, when interviewees were specifically asked questions to highlight agency on behalf of Latino/as, the majority of the responses, even among those individuals who exercised agency, were general, highlighting programs or policies that sought to benefit the entire student population. Thus, it is important to note that when

TABLE 11.1 Number of Interviews by Position and Institution

Institution	Faculty	Administrator/Staff	Total
Southwest Community College	4	3	7
West Coast College	3	2	5
Total # of Interviews by Role	7	5	12

TABLE 11.2 Participants by Institution, Position, Ethnicity, and Gender

Institution	Name	Title	Ethnicity	Gender
South West Community College	Jeffrey Carter	High Level Administrator	White	Male
	Brian Breslaw	Professor of Engineering	White	Male
	Andrea Guevara	Director of College Grants and Development	Hispanic/Latina	Female
	Brian Norwood	Chair of Physics and Engineering/Professor of Physics	White	Male
	Yolanda Velasco	Coordinator of Business Programs and Professor of Business	Hispanic/Latina	Female
	Maria Mendoza	Professor of Engineering and Counselor	Hispanic/Latina	Female
	Reynaldo Valle	Director of Admissions and Records	Hispanic/Latino	Male
West Coast Community College	Jeffrey Brooks	Dean of Student Services and Instruction	White	Male
	Amy Morris	Director of the Engineering & Industrial Technology/Physical Sciences and Professor of Engineering	White	Female
	Diane Torres-Richardson	Dean of the Office of Academic Affairs	Hispanic/Latina	Female
	Michael Fisher	Chair of Physical Science Department and Professor of Chemistry	White	Male
	Benjamin Rivera	Chair of Biology Department and Professor of Biology	Hispanic/Latino	Male

we highlight practitioner agency in our findings, in some cases it was difficult to differentiate whether agency was on behalf of Latinos/as specifically or *all* students.

Barriers to Agency

We found four kinds of barriers that limited participants' potential to exercise agency on behalf of Latinos/as in STEM. Two of the barriers were *structural*; one produced by the *one-sided transfer practices* of the four-year college and the other by the *data practices* of the community colleges. The other two barriers to practitioner agency had to do with *individual values and beliefs*: embracing an *ideology of color-blindness* and attributing Latino/a non-participation in STEM to a *family-centered culture*. These four barriers are described below.

Structural Barriers

One-Sided Transfer Articulation. The capacity of community colleges to transfer more Latinos/as in STEM fields is highly dependent on the willingness of the STEM faculty at the four-year college to accept credits earned within the major at the community college. The STEM transfer pathway, including who is eligible to transfer, how many transfers can be accommodated, and how many credits will be transferable are fully within the control of the four-year college. And, as Maria Mendoza, a professor of engineering and counselor, makes clear, faculty at the four-year college can change the rules of transfer suddenly and without consultation.

> Our main transfer institution makes changes in their engineering program and we don't find out about it at the department level until they've already published their plan. So we don't— they don't really—it's supposed to be a two-way articulation, but it really ends up one -way. And like recently this past year, you know, the intro to engineering class and the engineering graphics classes that are part of the first year for our students have been deleted out of the transfer plan, so now the students are saying, "Okay, I shouldn't be taking these classes." And our department is saying, "Yes, you should be taking them."

Clearly, the lack of curricular alignment in the requirements for the associate's degree in science and engineering and for transfer forces students described by Professor Mendoza forces students into the absurd situation of having to choose between earning the associate's degree and not meeting the transfer criteria, or foregoing the degree in order to qualify for transfer.

Professor Mendoza is an advocate for Latinos/as, but her efforts are undermined by the stratified structure of higher education and the perception among four-year college faculty of community colleges as academically inferior. Unless their programs are experiencing enrollment declines, faculty at the four-year college may not consider the consequences of their policies on students. The frustration and sense of powerlessness that can be heard in Professor Mendoza's description of SCC's relationship with the neighboring four-year college will likely sound familiar to community college faculty and administrators.

Data Practices. Data on the participation and performance of students in STEM majors that is routinely available disaggregated by race and ethnicity is an obvious way of raising awareness of underrepresentation or unequal outcomes. However, neither college seemed to exploit the power of data to communicate priorities and values consistent with their identity as HSIs. Respondents in the two colleges did not provide examples of strategic data use to call attention to Latinos/as' underrepresentation in STEM or to motivate greater support toward improving the performance of Latinos/as.

The Director of Admission at SCC, Reynaldo Valle, mentioned that although it could be done, disaggregating data on student transfer was not something the college did. Speaking about a recent effort to examine enrollment and graduation rates he said, "We do have our enrollment [data]. We can drill it down that way [by race and ethnicity]. I don't recall in looking at it, that we broke it down that way—we just had a flat graduation number." In response to a question about efforts to recruit Latinos/as into STEM majors, he said "Programs or initiatives in place for the sole purpose of recruiting Hispanic[2] students? No. I don't have an answer as to why we don't have programs—and we may I just—I don't know." He was not aware of any effort to "to try and raise our Hispanic population in this way [in STEM]. I know we have a lot of initiatives on recruiting and enrolling students, it just so happens that a lot them happen to be Hispanic. But to say that I know of a specific initiative that's targeted at that population—if there is, it's unbeknownst to me."

Individual Values and Beliefs

The Ideology of Color-Blindness. Some individuals, even among those who exercised agency on behalf of Latinos/as, seemed proud to claim color-blindness as a characteristic of the institution as evidence of their capacity to treat everyone equally. Ms. Guevara, a staff member, described the college and the town as "color-blind." She said,

> Well, we are surrounded by military bases, and when you enter the military you—any racism that you have brought with you gets literally beaten out of you by the time you've gotten through Basic Training, and by the time you have a lot of people of other colors and ethnicity to save your life and depend on you, you stop noticing what color people are … so it just—people don't notice as much what color anybody else is, and it's a very multi-racial society here …

For Brian Norwood, the Chair of Physics and Engineering at SCC, being color-blind was a legal issue. "Since we're not allowed to use that [race] as a factor in anything, it just makes sense [not to focus on any particular group]." He also thought that institutional practices work well or badly for everyone, regardless of their race or ethnicity. "The general method that works is gonna work for everybody." Professor Norwood's position here appears to be contradictory to an institution that is *Hispanic Serving*. However, his comment provides a reminder that having the HSI label is merely a designation and does not include a promise or goal of focusing on Hispanic students.

Similarly, Michael Fisher, a Professor of Chemistry, said, "Like I—I honestly don't look at my students—their heritage. In my head [I think], "Here's everybody. What can I do to keep you interested in what I'm doing or what I'm trying to [help you] learn?" Professor Fisher did not consider students in terms of their "heritage"; instead he focused on other differences. "When you look at who's in your class, it's already a very diverse group of people 'you are looking at.' For what it's worth, that diversity goes in several dimensions, not just what race or gender are you. It's how well prepared are you." When asked if he knew how Latinos and Latinas performed in his class, he said,

> If somebody wants to know, I could separate out the grades and say, "Here's how they're doing compared to everybody else." But I've never been asked to do that. I'm not aware of anybody trying to answer that kind of question. Many statistics I hear ... at least the ones I'm aware of, are basically just in general population how are they doing. It just doesn't have a label on it ... I don't know anyone else who does. I don't know if people go in and pull that information out to look at it.

Jeffrey Brooks, a Dean of Student Services and Instruction from WCC, also believed that practices worked in similar ways for all students.

> I don't know if there is any specific program that is designed, funded and supported by staff that intentionally targets any population ... I think in general, because we're open access, whoever comes to us, comes to us ... and we've reacted to whoever comes. And so there hasn't really been a need, or a perceived need in the STEM disciplines to focus on specific ethnic, cultural, diverse populations, other than the ones that come and then we say, "Okay, what is it that we need to do to serve these diverse populations?" So it's been more reactionary than intentional.

Family-Oriented Culture. Participants attributed Latino/a low levels of academic success to Latinos/as' closeness to their family and their unreasonable expectations for their children to work and help out financially. For instance, Professor Norwood associated Latinos/as' low participation in STEM with poor academic preparation, which he attributed to the family. "I don't know how much of this is peculiar to STEM but one of the biggest barriers is the lack of preparation ... I think that may impact more on Hispanics." Professor Norwood believed that Latino families expect their children to work and as a consequence they are not able to concentrate on school. "Because they [Latinos/as] have, I guess, a stronger family network. They are usually more heavily obligated to family work. So they come in already with an academic handicap. It's much easier for them to give up on it than it is for somebody else who doesn't have as much of a load on them already."

Amy Morris, an engineering instructor, perceived her students as having the potential to pursue majors in STEM but believed that Latinos/as' integration in STEM and participation in the kinds of activities and relationships that would expose them to in- and out-of-class opportunities were constrained by the family. "A lot of them isolate with their families and sometimes they work too much, which isn't gonna make them successful … sometimes they get a lot of pressure from their families and their families don't quite know how to support them, because families, a lot of times they need them to work to help support the family." "Latino students," Professor Morris said:

> get caught in this struggle between their families, which they love, because they're so family oriented and the careers that they're trying to establish. They are perfectly capable of doing [STEM], and in the long run, they'd be able to help their families for it, you know, if they could become engineers.

Additionally, Professor Morris thought that English might be an added barrier to Latinos/as' success. She mentioned that in the case of one of her "brightest students" it was obvious that English was not his first language.

> I think that they came from Mexico, and I met his family and his parents don't speak English, but he was so incredibly bright. And he really applied himself. So, I would say that there's a wide variety of skill levels. There are some that come in very under prepared, and maybe for whatever reason aren't able to get beyond that as quickly as others.

Actions that Reflect the Exercise of Agency

In this section we focus on actions that were motivated, at least in part, by individuals' awareness of Latinos/as' underrepresentation in STEM and a desire to create more opportunities for their success. Collectively, the analysis of interviews illustrated that individual agency is evident when practitioners take notice, unambiguously, of the constraints that challenge Latinos/as, and act proactively to address them. The individuals identified as exercising agency on behalf of Latinos/as are not heroic figures, they are ordinary faculty members, administrators, and staff who are moved to use their resources, knowledge, networks, and good will on behalf of students who they notice as needing support. The catalysts for exercising agency could be many, including professional norms, personal identification with underserved populations, departmental or institutional values, and critical consciousness of inequality.

Participants exercised two kinds of agency: Program Building and Self-Directed Remediation. Among the four individuals featured in this section, only one exercised agency to change his practices (Self-Directed Remediation), the other three created programs. The actions reflecting agency among Brian Breslaw, Amy Morris, Benjamin Rivera, and Yolanda Velasco are described below.

Program Building

Assisting Latinos/as to Become Scientists. The actions of Brian Breslaw, an engineering instructor, illustrate the exercise of agency to create new structures to socialize Latino/as into the community of scientists. Soon after arriving at SCC, Professor Breslaw noticed two contradictions. First, while he had many friends who were Hispanic, few of them were pursuing careers in engineering. Second, he observed that although his campus has a large Hispanic population, many Hispanic students do not go into engineering. Having noticed that the institution lacked structural supports for engineering students, he advocated for the establishment of programs to foster interest in the field of engineering. For example, learning from other states and institutions, he wrote a proposal for a grant to replicate a well-known minority serving engineer program, the first of its kind at SCC and the state. He also lobbied the president of his institution to provide space for a center that students could use to meet, study together, and interact with faculty. He acknowledged that creating resources for students is all about "personal involvement if you want to make something good happen."

Although Professor Breslaw was moved to act because he noticed that Hispanics were missing in his courses, it heeds mentioning that he also preferred to represent his motivations as color-blind. He described himself, first and foremost, as an advocate of engineering. He possessed a deep passion for the discipline and his identity was as an engineer, which he felt was far more important than race or ethnicity. He wanted to share his passion for engineering with others. To him science in general, and engineering more specifically, had nothing to do with race.

> The discipline for exact sciences, you know, that's required for engineering and exact sciences, I don't think that you distinguish so much Latinity from the rest. Maybe they are more passionate because it's their nature, but again, I am ... passionate myself. And I think I relate to that part [of the students' culture]. But because of the discipline required for these kinds of courses, the Latino identity takes a second stage because you are first an engineer and then you are a Latino ... I don't find the difference. I look at blacks, black engineers. I've worked with them. I don't see any difference from English or ... They are engineers.... So once you have a discipline in your brain, you don't act like a Latino or like an Anglo or like a black or like a Asian, you are an engineer in the first place.

Similar to Professor Breslaw, Professor Morris (first introduced in the section on "family oriented cultures") started a minority STEM program at WCC. Professor Morris started the program because she had noticed that Latinos/as in the Engineering Club, of which she is the director, "identified with each other a lot, and they started studying outside of class." She said that once the students started studying together and socializing they performed better in class.

> I noticed how much better they started doing. For example, before then, a lot of my students felt they just had to do it on their own. This one particular semester, it just seemed like this group connected. They socialized well together, they connected, and they started improving their grades.

She recognized that while Latino/a students benefitted from engagement, the institution did not adequately support this engagement, so like Professor Breslaw she wrote a grant application, was awarded funding, and used the funds to create a program in STEM for Latinos/as.

Developing College-Knowledge in the Latino Community. Yolanda Velasco, a professor and a Latina herself, felt that Latinos/as perform poorly because higher education does not understand the Latino/a community. "The administration doesn't mean to be neglectful, it's just that it doesn't understand our population. They don't understand the Hispanic population. They don't understand the dynamics in a Hispanic family." According to Professor Velasco, one of the primary ways that faculty and the institution can create a welcoming environment for Latino/a students and their family is by building trust, something that is "just neglected" by the institution. Professor Velasco believed that there is an inherent "mistrust of education institutions" in the Latino/a community and that colleges have to make a greater effort to reach out to the community, in their own language, and make them feel welcome. To familiarize the Latino/a community with higher education, Professor Velasco created an academy that educates parents about college admissions and financial aid.

Self-Directed Remediation of Teaching Practices. Professor Rivera, a biologist, acknowledged that he had expected students to know how to be successful, that "they should know how to study, how to read." "But," he said, "I realized they really didn't know how to do much of this stuff." Professor Rivera decided that it was his responsibility to teach students the course content as well as the skills to learn the content.

Professor Rivera took the time to observe his students' behaviors and adapted his practices to better meet their needs. For example, he described an instance when he asked a student if he could look at his notes and was surprised to see that during a 50-minute lecture on evolution, the student had only written a couple of sentences. He realized that this was not an isolated case, but a pattern he saw across all of his classes. The students, he said, "haven't learned how to take notes, how exactly to study, how to prepare for courses, the basic study skills."

Professor Rivera came to his position with the mindset that his job was to teach biology, but if he wanted his students to be successful they had to learn how to learn biology. He altered his practices to "teach them study skills." "Instead of actually doing a biology lab," he said, "I take one lab period and teach the students how to use a textbook." Professor Rivera has also added study skill enhancing assignments to his syllabus to help students learn basic study skills. One assignment he added to the content of the course required students to take an online quiz focused on teaching students their learning style, "So are they visual? Are they kinesthetic? Are they auditory?" Then, Professor Rivera tells students to remember their learning style "for future use." When students have questions about their performance on biology exams, he tells them to come to his office hours and to "bring that little sheet" with their learning style results. During their meeting the conversation will go something like this:

> Professor Rivera: "Okay, now let's take a look at what it said. Now how did you study?"
>
> Student: "Well I used flashcards or I rewrote my notes."
>
> Professor Rivera: "Well let's take a look. Was that a good way to study for you?"

Professor Rivera also added an assignment to his syllabus regarding visiting the tutoring center. He acknowledged that "the difficulty is getting them there [tutoring center]." He has found that "either they don't know about it or if they do know about it, they think "well only dumb people go to tutoring centers." He created an assignment to teach the students where the tutoring center is located and the resources it offers. The student has to find "out who is teaching or who is tutoring this class and what is their schedule." He believed that an assignment like this helps get students "in the room" and also helps students see the value of using the center.

Finally, Professor Rivera, like Professor Velasco, also took the time to involve his students' parents, so that they would have a better understanding of what college is and why their support is important. "I take my students but I also tell my students, 'Bring your parents. Bring your family. Bring your nieces. Bring your nephews. Bring everybody. Bring half of the city with you. I don't care.'" "When they invite their parents their parents go with them and then they kind of take a look at the stuff that they're doing, they take a look at the worksheets, they take a look at the stuff that they're learning and they're saying, 'Oh so there is a purpose to this.'"

Discussion

Table 11.3 below summarizes the findings based on the following characteristics of the participants' responses: Awareness of Latinos/as' participation in STEM (columns 1 and 2); Awareness of constraints and their effect on Latinos/as' STEM participation (column 3); Color-blindness (column 4); Latinos/as' participation in STEM attributed to personal and cultural factors (column 5); Latinos/as' participation in STEM attributed to institutional factors (column 6); and Exercised agency to improve Latinos/as' participation and success in STEM (column 7).

TABLE 11.3 Findings Based on Participant Responses

	(1) Was not aware of Latino participation in STEM	(2) Aware of Latino participation in STEM	(3) Awareness of constraints and their effect on Latino STEM participation and success	(4) Color-blindness	(5) Latino underrepresentation in STEM attributed to individual and cultural characteristics	(6) Latino underrepresentation in STEM attributed to institutional factors	(7) Acted to improve Latino participation and success in STEM
Reynaldo Valle Admissions Dean	Did not know if among the cc's many initiatives any were focused on Latinos/as						
Andrea Guevara Director, Administration and Operations				Described the institution's geographic location as multi-racial; therefore people's "color" not noticed.			
Brian Norwood STEM Department Chair				Invoked legal restrictions preventing focus on race. Asserted that practices work well or badly regardless of race or ethnicity.	Lack of preparation impacts Latinos more than other groups; Latino students expected to work; come with academic handicap; easier for them to give up than students with fewer responsibilities.		

Michael Fisher STEM Instructor	Looks at everyone the same regardless of race. Focuses on differences in academic preparation (not race).	
Jeffrey Brooks STEM Dean	CC is open access and serves everyone.	Not aware of programs specifically for Latinos within his own campus.
Amy Morris STEM Instructor	Get a lot of pressure from family to work; family does not know how to support them; family needs them to work to help out. Some very underprepared and not able to catch up as quickly as others.	Noticed that Latinos in her classes studied together and seemed to do better when they worked together. Institution did not provide support for community building among Latino students.	Applied for grants and started a special program based on a national model.

TABLE 11.3 Continued

(1)	(2)	(3)	(4)	(5)	(6)	(7)
Maria Mendoza STEM Student Support Staff		Articulation with 4-year is one-sided and can change from year-to-year without consultation. Requirements for AAS and Transfer are not aligned forcing students to make choices that are irrational.				
Brian Breslaw STEM Chair			Racial identity not as important as professional identity.	Noticed few Hispanics majoring in STEM, even though they represent a sizable proportion of the student body.	Noticed lack of institutional supports for STEM students in general.	Identified types of support provided to Latinos in STEM in other institutions. Wrote proposal and acquired grant to replicate national model for Hispanics in STEM. Secured space to provide Hispanics in STEM a place to gather.

Yolanda Velasco Instructor and Administrator in a non-STEM area	College does not understand the Latino community, including their mistrust of government. College needs to reach out to Latinos in their own language.	Created academy to familiarize Latino parents with college admissions, financial aid, etc.
Benjamin Rivera STEM Instructor	Noticed a pattern across his classes: students lacked the knowledge and experience to study effectively. Recognized that the fear of being labeled "dumb" kept students from using the tutoring center.	Changed teaching practices to meet students' needs; taught them how to read biology textbook and how they learn best. Created an assignment that required students to visit the tutoring center. Invited parents on field trips to learn what their children do in college.

Among the six participants who held academic positions in STEM, just three, Brian Breslaw (STEM Instructor and Chair), Amy Morris (STEM Instructor and Chair), and Benjamin Rivera (STEM Instructor) took notice of institutional and classroom characteristics that disadvantaged Latinos/as' successful participation in STEM. Professors Breslaw and Morris perceived that their respective colleges were not providing sufficient support and took the initiative to find the resources to replicate a well-known academic and support program that was originally created to increase Latinos/as participation in STEM. Both also found ways to create a physical location where Latinos/as taking courses in STEM could gather, study together, meet with faculty members, and plan their participation in national conferences for minority students in STEM. At SCC, the space was so popular that students who had transferred to the four-year college continued to use it as a home base because the four-year college did not have anything like it.

While Professors Breslaw and Morris invested their time in acquiring external funds to expand academic support as well as dedicated counseling and advising services, Professor Rivera's in-classroom observations of students' difficulties with the basic skills of note taking, studying for the test, and seeking help led him to recognize that his expectations for students who know how to be biology students were not going to be met. Rather than throwing his hands up in the air and attributing failure to under preparedness and wishing for better students, he remedied his practices. That is, he scaffolded learning activities based on his observations of students' skills (e.g., how to take notes as well as his intuition and anticipating that students might avoid the tutoring center for fear of appearing to be dumb). Rather than exhorting students to take advantage of the tutoring center, he created an assignment that gave them a specific purpose for their visit and a script for what to do once there. Making a visit to a tutoring center a required assignment for a biology class may not seem all that important. We are highlighting it because throughout much of the work the Center for Urban Education conducts on college campuses of all types, we frequently hear comments like, "we have great resources, but they [minority students] don't use them" or more crudely, "you can take the horse to water but you can't make it drink." It is difficult for institutional actors who have not themselves experienced difficulties or who assume that academic success is an individual responsibility to consider that the underutilization of resources is related not to lack of motivation but the fear of confirming stereotypical expectations others have of minority students, or the fear of not knowing what to do, what to say, how to behave. For those who have learned to expect support and feel entitled to it these fears may seem illogical. But the trepidation that Latinos/as might feel about entering the academic support center is not unwarranted as academic center staff, faculty, as well as peer tutors may not be conscious of whether their practices facilitate or inhibit help-seeking.

The other three participants with academic appointments shared the view that race or ethnicity is not consequential and essentially that practices within

STEM (or generally in an academic setting) are impervious to students' race or ethnicity. Professor Fisher said he focused on the students' academic preparation; Professor Norwood thought that focusing on Latinos/as had legal implications; and Dean Brooks reasoned that in being open access institutions community colleges always care about how to serve the diverse needs of the many types of students they attract. Professor Norwood thought that Latinos/as' came from a culture where there is an expectation that the have a job while going to school and, as a consequence, they accumulate "academic handicaps" which prevent them from being successful. He also thought that work responsibilities made it easier for Latinos/as to "give up." Like Professor Norwood, Professor Morris also spoke about the burden of work for Latino/a students and their under preparation as circumstances of their lives; however, unlike Professor Norwood, she did not focus on these personal factors as a justification for their underrepresentation in STEM. She, along with Professors Breslaw and Rivera, were aware of the personal challenges faced by the students and they concentrated on what they could control—building and shaping academic environments that could enable persistence, success, and transfer in STEM despite the challenges in Latino/a students' personal circumstances.

Being blind to race and ethnicity, in particular being unconscious of Latinos/as' representation and participation in STEM programs, seems incongruous in colleges that have been designated as Hispanic Serving. Professors Norwood, Fisher, and Dean Brooks beliefs that race is absent from practices, unfortunately, prevents them from investing their power and resources in the creation of opportunities for Latinos/as. Unintendedly, their belief in color-blindness and confidence in the neutrality of college structures, policies, and practices blinds them to the added burden that their stance creates for Latinos/as.

In addition, color-blindness among powerful individuals, such as Professor Norwood and Dean Brooks, undermines institutional capacity to engage in equitable practices. For example, the value placed on color-blindness, particularly among academic leaders whose actions signal what is important, lessens the possibility that staff members will recognize the need to act purposefully in regards to Latinos/as. For example, Mr. Valle, the director of admissions, did not perceive that the recruitment of Latinos/as for STEM programs was an institutional priority.

Color-blindness makes it more difficult to notice institutional practices that hide inequalities. Ms. Guevara noted that although her campus had the data capacity to disaggregate by race and to track transfers, they did not routinely do so. The barrier to knowing how Latinos/as were faring was not data availability, but the omission by campus leaders of Hispanic Serving colleges of routinely monitoring how the college is performing in regards to Latinos/as. Clearly, the prevalence of color-blindness makes the good efforts of Professors Breslaw, Morris, and Rivera more difficult and reduces their impact to a small number of students.

Conclusion

The potential for agency (i.e., doing the good) exists in every professional; however, the findings we report here and elsewhere show that only a small number of professionals are identifiable as institutional agents for Latinos/as in STEM. The actions and beliefs of the 12 faculty members, administrative leaders, and staff that we have described raise an important question: Assuming that individuals like Professors Breslaw, Morris, and Rivera contribute in measurable ways to the success of Latinos/as in STEM, what circumstances could be engendered so that there would be more of them?

In order to address this question we turn to Seo and Creed (2002) whose dialectical framework on institutional change considers how individuals embedded in their own organizational settings develop the agency to transform them. They confront the longstanding paradox in institutional theory that if institutions are, by definition, firmly rooted in taken-for-granted rules, norms, and routines, and if those institutions are so powerful that organizations and individuals are apt to automatically conform to them, then how are new institutions created or existing ones changed over time (DiMaggio & Powell, 1991; Seo & Creed, 2002)? The dialectical framework positions institutional change as a result of *institutional contradictions* and *human praxis*. As defined by Benson (1977), contradictions are incompatibilities or tensions in the organization and praxis is the practice of becoming active agents in the reconstruction of social relationships (Benson, 1977). This concept of praxis is defined as a particular type of human action, situated in a given sociohistorical context but driven by inevitable contradictions (Benson, 1977). Praxis operates as a mediator between institutional contradictions and institutional change. Drawing on Benson's (1977) dialectical perspective, the framework focuses on how various inconsistencies or tensions (contradictions) transform the embedded actors into institutional agents and how the contradictions foster subsequent change processes. Although practitioners can become reflective at any time, the likelihood of a shift in awareness that can change practitioners from passive participants into change agents increases when practitioners continually experience contradictions in a given context.

The individuals in this study who displayed actions that are characteristic of "exercising agency" on behalf of Latino/as in STEM appeared to be motivated to act by contradictions within their settings. For example, Professor Breslaw noticed the underrepresentation of Latino/as in STEM in a college where they are the majority. Professor Rivera started out with the expectation that students would have the skills to take notes, to study effectively, and to comprehend the biology textbook. His expectations were not realized, which caused him to alter his vision of what it means to be an instructor, not just an instructor of content but of study skills as well. Professor Morris noticed that her students succeeded when they had a place to engage, yet no place existed on her campus.

These contradictions were catalysts to the actions they took to improve opportunities for Latinos/as in STEM.

To develop practitioners to exercise agency and act as institutional agents, administrators and other campus leadership can enact practices that foreground contradictions. Administrators can set the stage for awareness of contradictions by instilling a culture based on race consciousness, accountability for racial equity in student success, and a professional ethic based on doing the good for students.

Notes

1. National Science Foundation (Grant No. 0653280). Any opinions, findings, and conclusions or recommendations expressed in this material are those of the author(s) and do not necessarily reflect the views of the National Science Foundation.
2. We use the terms *Latinos* and *Latinas* throughout the chapter; however, when quoting from participants' interviews we use their language. Mr. Valle and several others used the term *Hispanic*.

References

Bensimon, E. M. (2007). The underestimated significance of practitioner knowledge in the scholarship of student success. *The Review of Higher Education, 30*(4), 441–469.

Bensimon, E. M., & Dowd, A. C. (2009). Dimensions of the transfer choice gap: Experiences of Latina and Latino students who navigated transfer pathways. *Harvard Educational Review, 79*(4), 632.

Bensimon, E. M., & Dowd, A. C. (2012). *Developing the capacity of faculty to become institutional agents for Latinos in STEM*. Los Angeles: University of Southern California.

Benson, J. K. (1977). Organizations: A dialectical view. *Adminstrative Science Quarterly, 22*, 1–21.

Cabrera, A. F., Burkum, K. R., & La Nasa, S. M. (2005). Pathways to a four-year degree: Determinants of transfer and degree completion. In A. Seidman (Ed.), *Student Retention: Formula for Success* (pp. 155–214). New York, NY: Rowman & Littlefield.

Cabrera, A. F., La Nasa, S. M., & Burkum, K. R. (2001). *Pathways to a four-year degree: The higher education story of one generation*. University Park, PA: Center for the Study of Higher Education, Penn State.

Chang, M. J., Cerna, O. S., Ham, J., & Sa'enz, V. (2008). The contradictory roles of institutional status in retaining underrepresented minorities in biomedical and behavioral science majors. *The Review of Higher Education, 31*(4), 433–464.

DiMaggio, P. J., & Powell, W. W. (1991). Introduction. In W. W. Powell & P. J. DiMaggio (Eds.), *The new institutionalism in organizational analysis* (pp. 1–38). Chicago, IL: University of Chicago Press.

Freire, P. (1970). *Pedaogy of the oppressed* (M. B. Ramos, Trans.). New York, NY: Continuum.

Garcia, G., & Hurtado, S. (2011, April). *Predicting Latina/o STEM persistence at HSIs and non-HSIs*. Paper presented at the Annual Meeting of the American Educational Research Association, New Orleans, LA.

Hurtado, S., Cabrera, N., Lin, M., Arellano, L., & Espinosa, L. (2009). Diversifying science: Underrepresented student experiences in structured research programs. *Research in Higher Education, 50*(2), 189–214.

Patton, M. Q. (1990). *Qualitative evaluation and research methods* (2nd ed.). Newbury Park, CA: Sage.

Rendón, L. (1994). Validating culturally diverse students: Towards a new model of learning and student development. *Innovative Higher Education, 19*(1), 33–51.

Seo, M., & Creed, W. E. (2002). Institutional contradictions, praxis, and institutional change: A dialectical perspective. *The Academy of Management Review, 27*(2), 222–247.

Stanton-Salazar, R. D. (1997). A social capital framework for understanding the socialization of racial minority children and youths. *Harvard Educational Review, 67*(1), 1–40.

Stanton-Salazar, R. D. (2001). *Manufacturing hope and despair: The school and kin support networks of U.S.-Mexican youth.* New York, NY: Teachers College Press.

Stanton-Salazar, R. D. (2011). A social capital framework for the study of institutional agents and of the empowerment of low-status youth. *Youth & Society, 43*(3), 1066–1109.

Tsui, L. (2007). Effective strategies to increase diversity in STEM fields: A review of the research literature. *The Journal of Negro Education, 76*(4), 555–581.

12

ACHIEVING SUCCESS

A Model of Success for Black Males in STEM at Community Colleges

Robert T. Palmer and Zachary M. DuBord

Improving college access and success among Black males has garnered tremendous attention. For example, there have been policy reports from the College Board Advocacy and Policy Center and the Center for the Study of Race and Equity in Education. Furthermore, there are a number of journals, such as *Journal of African American Males in Education, Challenge Journal: A Journal of Research on African American Men, Spectrum: A Journal on Black Males*, among others devoted to scholarly inquiry and providing insight into the experiences, conditions, and challenges facing Black males in education and beyond. At the same time, many educational researchers have sought to provide critical insight and raised thought provoking questions about Black males through peer reviewed articles and books.

Indeed, this focus on Black males is warranted. Black male students currently account for 4.3% of the total enrollment at four year postsecondary institutions in the United States (Strayhorn, 2010). Incidentally, the percentage of Black males who are enrolled in college is the same as it was in 1976 (Strayhorn, 2010). The uneven participation of Black males in higher education has caused researchers to take note of a growing gender imbalance between Black males and females enrolling and succeeding in higher education (Palmer & Maramba, 2012).

The college enrollment of Black males has a significant impact on their participation in science, technology, engineering, and mathematics (STEM). Interestingly, though not disaggregated by gender, research shows that in 2006, 34% of Black freshmen planned to major in STEM. Ironically, the percentage of Black students who expressed an interest in STEM exceeded White freshmen by 29.5% (National Science Foundation [NSF], 2008]). Despite this, few Blacks actually graduate with baccalaureate degrees in STEM. Specifically, while there has been a modest increase in the number of Blacks with bachelor's degrees in science and engineering, the number of Blacks with degrees in STEM has yet to reach parity with their representation in the general population (NSF, 2006).

Indeed, the disproportionate number of Blacks with degrees in STEM is problematic because the U.S. Census Bureau predicts that racial minorities (e.g., Blacks, Latino/as, Asian Americans, and Native Americans) are expected to grow rapidly over the next few decades (Museus, Palmer, Davis, & Maramba, 2011). Consequently, researchers, policymakers, and educational leaders have strongly emphasized the importance of growing the academic talent of this population so they can participate in STEM fields (Harper & Newman, 2010; Museus et al., 2011; Palmer, Maramba, & Dancy, 2011; Perna et al., 2009). In fact, scholars have argued that the global competitiveness and economic vitality of the United States will be impaired if little is done to increase the participation and success of minority students in STEM (Harper & Newman, 2010; Museus et al., 2011).

Community colleges have played an essential role in serving as a gateway for increasing the number of minorities in STEM (Hagedorn & Purnamasari, 2012; Starobin, Laanan, & Burger, 2010; Starobin, Jackson, & Laanan, 2013). Despite the fact that community colleges serve as an entry point for higher education for many Black males (Bush & Bush, 2005; Wood, 2011; Wood & Turner, 2011), a large number of Black males who begin their postsecondary through community college do not persist to graduation or transfer to four-year institutions (Bush & Bush, 2005; Hagedorn, Maxwell, & Hampton, 2001-2002; Wood, 2011; Wood & Turner, 2011). Indeed, Black males are at an increased likelihood of prematurely departing from community college compared to males of other racial and ethnic groups (Hagedorn et al., 2001-2002).

Evidence from Esters and Mosby (2007) documents the dismal participation of Black males in community colleges. Specifically, using data from the Integrated Postsecondary Education Data System (IPEDS), Esters and Mosby (2007) found that Black males have the lowest graduation rates compared to males from other racial/ethnic groups, with only 16% graduating in a three-year time span. Wood (2011) argued that Black males have the lowest mean grade point average (GPA) of men in community colleges. Using 2006 data from the U.S Department of Education, Wood (2011) explained that White, Latino, and Asian American males had an average GPA of 2.90, 2.75, and 2.84 whereas Black males had a GPA of 2.64.

Purpose of Chapter

Indeed, community colleges play a critical role in serving as a gateway to STEM for minority students. Nevertheless, as noted, many Black males who enter community colleges leave without attaining a degree or do not transfer to a four-year institution to continue their education. Given this, this chapter will draw from extant literature on Black male collegians, STEM, and community colleges to construct a model that could be useful to increase the success of Black males in community colleges pursuing training for STEM. This study makes a significant contribution to the literature because there is a scarcity of studies that provide salient information for practitioners, faculty, and administrators in community college to increase the success of this population in STEM at two-year colleges.

Academic Preparedness

One of the critical issues facing Black males as they transition to higher education is lack of academic preparedness. There are multiple factors in K-12 that contribute to the lack of preparedness among Black males. For example, research show that school funding is related to the quality of education that students receive (Hagedorn & Purnamasari, 2012; Museus et al., 2011). School funding is derived from local property taxes, and, as White students live in more affluent neighborhoods, they are more likely to attend schools that receive more funding per pupil than schools in less wealthy communities where many Blacks, Hispanics, and low-income students live (Flores, 2007; Hagedorn & Purnamasari, 2012). The funding disparity between school districts is correlated with the kind of resources schools can provide for students to facilitate student learning (Hagedorn & Purnamasari, 2012). Given that minority students are disproportionately more likely to attend schools that are funded unequally, students at these schools are at a disadvantage because they are placed in larger classes and lack access to the latest books, laboratories, instructional material, and technology—all of which impinge upon the quality of their learning (Museus et al., 2011).

Similarly, the systematic tracking of Black males into lower academic tracks during K-12 is another critical factor contributing to their lack of academic preparedness as they transition into postsecondary education (Hagedorn & Purnamasari, 2012). Academic tracking promotes racial and ethnic inequality because students who are placed in high-achieving academic tracks are exposed to more complex and challenging classroom instruction than those who are placed in low-achieving academic tracks (Museus et al., 2011). Empirical research shows that Blacks are overrepresented in low-ability or remedial tracks (Hagedorn & Purnamasari, 2012). Indeed, while Blacks are overrepresented in lower academic tracks, they are underrepresented in gifted education and advanced placement courses (Hagedorn & Purnamasari, 2012). In some cases, the inner city schools that Black students attend do not offer advanced placement courses (Lewis, 2003).

The prevalence of unqualified teachers in the K-12 system has been cited as another reason for the lack of academic preparedness among Black students

(Hagedorn & Purnamasari, 2012; Museus et al., 2011). A report from the NSF (2010) underscored the severity of racial minority students not having equal access to qualified teachers in K-12. For example, the report noted that, in 2004, White fifth graders were 51% more likely to be taught by teachers with a master's or advanced degree than their Black and Hispanic peers. Similarly, Flores (2007) and others (e.g., Hagedorn & Purnamasari, 2012; Museus et al., 2011) have explained that students attending predominantly Black and Hispanic schools are twice as likely to be taught by teachers with three years of teaching experience or less, compared to those attending predominantly White schools.

Finally, teacher expectations have been noted as a salient factor responsible for the dearth of academic preparedness among Black males as they transition to postsecondary education (Museus et al., 2011). In particular, research has noted how low teacher expectations can hinder the achievement of racial minority students in math and science courses (Museus et al., 2011). Indeed, the relationship between teacher expectations and academic achievement appears to be a reciprocal one. That is, while teacher expectations influence academic achievement, students' academic performance can also affect teachers' expectations of those pupils. Teachers may be more likely to develop expectations about and treat their students in a manner that is more consistent with those students' performance on standardized assessments than their actual abilities (Museus et al., 2011). Thus, given that racial minority students are likely to perform lower on standardized math and science examinations than their majority counterparts, teachers are more likely to have higher expectations for White than minority students (Hagedorn & Purnamasari, 2012).

In turn, research demonstrates that teacher expectations can influence academic performance, suggesting that those expectations can become a self-fulfilling prophecy for students. Mathematics and science courses are viewed as higher-order disciplines, and teachers are inclined to perceive racial and ethnic minority students as lacking ability in those areas and send subtle messages that such disciplines are White male domains (Museus et al., 2011). Moreover, such messages can lead to differences in teaching behavior and subsequent achievement (Museus et al., 2011).

One of the ways that community colleges have responded to academic preparedness among their students is offering developmental or remedial academic coursework (Perin, 2005). The purpose of college remediation is to enhance academic deficiencies for students scoring below proficiency on standardized admissions testing and college placement testing. Generally with remediation, students are provided academic support, with program components ranging from a single course offering to more comprehensive academic and social support services, such as tutorial support, counseling, and study seminars. Research shows that community colleges are open access institutions that admit a wide variety of students (e.g., minority students, part-timers, adult returners, exceptional or special needs, and veterans), many of whom are in need of remediation services (Nevarez & Wood, 2010), particularly low-income minority students (Hagedorn et al., 2001-2002; Hagedorn, Cabrera, & Prather, 2010).

Research indicates that a large number of Americans leave high school underprepared for the academic demands of higher education. Nevertheless, Black students are twice as likely as White students to be in need of remedial services because of their experiences in K-12. Indeed, sufficient academic preparedness is critical to access and success in STEM majors during postsecondary education (Maton, Hrabowski, & Schmitt, 2000; Museus et al., 2011).

Financial Support

Aside from preparedness, many Black students may need financial support to help offset the cost of tuition and are perhaps working while attending school, which can detract from academic success. Indeed, researchers (e.g., Pascarella & Terenzini, 2005) have noted a link between financing college and persistence. According to Fenske, Porter, and DuBrock (2000), the availability of financial aid is among the top five factors related to the persistence of minority students in STEM majors.

One factor that has exacerbated college students' ability to finance college is declining state government financial support of public higher education. Paulsen and St. John (2002) explained that during the last few decades, the federal government has shifted from using grants as the primary means to facilitate college attendance to using loans. In fact, they characterized the last few decades "... as a period of high tuition, high aid, but with an emphasis on loans rather than grants" (p. 195). They argued that this shift in federal student aid policy has been more problematic to low-income students compared to more affluent students and more burdensome for Black students than their White counterparts. Interestingly, a quantitative study that Leppel (2002) conducted on the similarities and differences in the college persistence between men and women emphasized that in addition to relying on loans to finance college, students feel the need to work more, particularly off campus, to help finance their college degree. Indeed, many community college students have many external commitments, such as the need to work off campus, which can be a liability rather than an asset to their retention and persistence (Tinto, 1993).

In addition, although community colleges have low tuition compared to their four-year counterparts, the declining state support for higher education has shifted more middle and upper middle-class students to community colleges (Rhoades, 2012). "Indeed, this, in turn, increases competition for seats in community college classroom at a time when funding for community colleges is being slashed and fees are increasing" (Rhoades, 2012, para. 4). As community colleges attract more students from affluent backgrounds, "opportunity is being rationed and lower-income students (many of whom are students of color) are being denied access to higher education" (Rhoades, 2012, para.

4). If educators, administrators, and policymakers aim to increase the number of Black males pursuing STEM majors in community colleges, considerations must be given to reducing financial hardships among minority students and stop denying them access to postsecondary education.

Faculty Support, Institutional Climate, Pedagogical Approaches, Psychological Indicators, Supportive Family, and Quality and Quantity of External Commitments

Research indicates that faculty plays a vital role in the success of minority students in STEM (Pascarella & Terenzini, 2005). While research has supported this connection, some research has suggested that minority students lack supportive relationships with White faculty in postsecondary education (Guiffrida, 2005). Furthermore, research illustrates that the absence of minority faculty who can function as role models for minority students, faculty with insufficient preparation, and faculty who maintain low expectations for minority students can hinder their success in college (Museus et al., 2011). But, faculty interaction, support, and encouragement have all been linked to minority students' success in STEM (Museus et al., 2011).

Indeed, while Black students in community colleges have different experiences than their counterparts in four-year institutions (Flowers, 2006), research has shown that positive faculty-student interaction is critical to student success in community colleges. In fact, McArthur (2005) argued, "faculty members represent the authority figure, the mentor, and the role model that may not appear anywhere else in the student's life. Because the faculty members are in such a position, their influence over students can be very significant" (para. 4). Given the importance of faculty-student relationships in community colleges, research has provided context about factors that will facilitate positive faculty-student interactions with educators in community colleges and Black students. For example, from the voices of 28 Black male community college students, Wood and Turner (2011) explained that four factors were salient to positive faculty-student engagement for Black male community students. Specifically, they noted that faculty should present themselves as friendly to help students to feel comfortable. They also explained that faculty should monitor students' progress and display concern for their performance in class. Similarly, the authors indicated faculty should listen intently to students' concerns. Furthermore, Wood and Turner explained that faculty should encourage the success of Black males by reinforcing the importance of striving for academic achievement.

Other researchers have echoed Wood and Turner's (2011) finding about the characteristics faculty should display to be viewed as supportive by Black students. Specifically, studies by Bush (2004) and Jordan (2008) found that it is important for faculty to be attentive to students' holistically by listening to their experiences, perceptions, and contributions. Beckles (2008) emphasized that doing so can provide a safe space, which will allow students to feel comfortable discussing critical issues and concerns with faculty. Indeed, while research has provided some indication about essential characteristics of faculty-student interaction and Black male success in community college, little, if any research, has explored faculty-student engagement and success in STEM majors for Black students in community colleges. Therefore, it is significant that research provide more contexts between these dynamics.

To our knowledge, little research has discussed how minority students perceive the institutional climate of community colleges. Museus et al. (2011) explained that a negative institutional and classroom climate can be detrimental to the academic success of minority students generally and to those in STEM specifically. For example, research indicates that minority students experience alienation, marginalization, and isolation on the campuses and in the classroom at predominantly White institutions (Gonzalez, 2003). Conversely, institutions that have more collective orientations and humanized environments may positively impact the success of minority students at two- and four-year institutions (Museus et al., 2011).

In addition to a supportive climate, the pedagogical approaches of faculty can have an important influence on minority student success in STEM (e.g., Museus et al., 2011; Seymour & Hewitt, 1997). In particular, four pedagogical practices have been linked to positive outcomes among minority students in STEM: (a) small and interactive classrooms, (b) collaborative learning, (c) a diverse and culturally responsive curriculum, and (d) a curriculum relevant to real-world problems. Historically, STEM courses have been based on large lecture type courses, and such courses have a negative influence on the experiences of minority students (Museus et al., 2011). Conversely, smaller and more interactive classes have been associated with success among racial and ethnic minority students in STEM (Perna et al., 2009). Second, researchers have found collaborative learning to significantly influence professional competencies, development, and analytical thinking skills, and success (Cabrera, Crissman, Bernal, Nora, & Pascarella, 2002; Museus et al., 2011). Third, some scholars have discussed the lack of cultural relevance in the STEM curriculum and need for faculty to address this reality (Perna et al., 2009) and the absence of STEM curricula that is pertinent to real-world issues. Thus, a curriculum that is both relevant to the cultural backgrounds and real-world problems has been noted as critical to increasing the success of minority students in STEM (Museus et al., 2011). Indeed, while curriculum matters to the success of minority students in STEM, the pedagogical approaches of faculty who teach STEM related courses in community college are unclear.

In addition to pedagogy, faculty (and academic advisor—who will be discussed in the subsequent section) can play a central role in helping to increase psychological factors (e.g., self-concept and self-efficacy) that are linked to student achievement in STEM (Museus et al., 2011). Higher levels of self-concept and self-efficacy are associated with greater likelihood of entering STEM fields, commitment to science and math, and higher levels of adjustment, performance, and success minority students in STEM (Museus et al. 2011). Indeed, being attentive to psychological factors is vital because research illustrates that Black male community college students without self-confidence are at an increased likelihood of attrition (Hagedorn et al., 2010). Thus, improving psychological factors among Black males in community colleges may lead to positive educational outcomes.

In the community college context, consideration must not only be given to providing a supportive campus community via faculty and other staff agents, but increased attention must be placed on students' interaction with family and connections with external commitments in order to help bolster success among Black males in STEM (Flowers, 2006; Wood, 2011). Because few community colleges have residence halls, community college students may be more connected to their families and more engaged in external obligations (Tinto, 1993). Depending on the quality of these familial relationships and both the kinds and depth of these external obligations, these factors can either have a positive or negative impact on community college students' success in STEM related coursework.

Faculty–Student Engagement in Research Opportunities

Indeed, a strong body of research has documented the importance for faculty–student engagement in research to enhance persistence among students in STEM in four-year institutions. According to Palmer et al. (2011), not only does participating in research opportunities with faculty help attract and retain students in science, but it also enhances the educational experiences of students, and serves as a linchpin to careers in science. Additionally, student engagement in research with faculty has been shown to enhance knowledge and understanding of STEM disciplines, facilitate faculty–student interaction, foster problem-solving, technical and presentation skills, engender self-confidence, and provide greater insight and clarification of career goals (Palmer et al., 2011).

Limited investigation has focused on faculty–student engagement in research and outcomes among community college students in STEM. This lack of research is not an indicator of the value that community colleges place on the importance of faculty–student engagement in research activities. For example, some community colleges are members of the Mathematics Engineering Science Achievement (MESA), which has the MESA Community College Program (MCCP). MCCP is a NSF-funded program designed to increase the number of low-income minority students in engineering who transfer to four-year colleges (Starobin et al., 2013).

According to Starobin and colleagues (2013), Highline Community College, an affiliate with MESA, uses funding from MCCP to provide classroom space for academic and social activities for STEM students, tutoring services, informal communication with a faculty advisor, and participation in the annual Human Powered Paper Vehicle (HPPV) competition, which is an activity that promotes academic and social interaction among students, mentors, faculty, and support staff. It is unknown how many community colleges are affiliated with MESA and the extent to which this program facilitate student-faculty engagement in research. Nonetheless, this program illustrates that community colleges recognize the significance of faculty working collectively with students on research to help increase their self-efficacy and likelihood of transferring to STEM majors at four-year institutions.

Academic Advisors

Research has shown a relationship between academic advising and persistence (DeSousa, 2005). Some research has shown a small, but negative relationship between persistence for students in STEM and advisors (Grandy, 1998). While Museus et al. (2011) noted that while the impact of academic advising on the success of minority students in STEM is unclear, evidence suggests that poor advising can negatively influence persistence in STEM and cause students to prematurely depart from STEM (Maton & Hrabowski, 2004).

Indeed, in community colleges the role of an academic advisor is critical to student success (McArthur, 2005). McArthur (2005) emphasized that academic advising is an essential factor to students for community college students. Echoing this sentiment, King (1993) observed that inadequate academic advising is linked to negative outcomes among community college students while quality advising positively impacts retention and persistence. Bahr (2008) stated that academic advising has the potential to be particularly beneficial to community college students who are academically deficient. Not only is advising important for student success in community colleges, King (1993) urged that academic advisors play a vital role in helping students transfer to four-year institutions (Hagedorn et al., 2010; Malcom, 2010). Malcom (2010) argued that academic advisors are critical in the transfer process because they are aware of barriers community college students are likely to encounter as they transfer and can provide students with information to increase their success at four-year institutions. Again, while academic advising is a salient component of student success, little research has examined the role that advising plays in the success of students pursuing STEM majors in community colleges.

Peer Support

In his theory of student departure, Tinto (1993) emphasized the importance of students becoming academically and socially integrated into the campus. Academic integration involves experiences that formally pertain to students' education and connect students with faculty and staff (Palmer et al., 2011). On the other hand, social integration entails formal and informal experiences that facilitate students' interactions with their peers, faculty, and staff and that occur primarily outside the academic purview of the institution (Tinto, 1993). Indeed, Tinto's theory primarily considers the context of four-year institutions as students who attend community colleges may find it difficult to become academically and socially into the institution (Tinto, 1997). Nevertheless, in a study that Tinto (1997) conducted at Seattle Central Community College, he delineated how a learning community was used to help students become integrated into community colleges. Specifically, Tinto (1997) explained that the learning community via classroom interactions linked students to peers, who served as a basis of support; bridged the divide between academic and social issues in and outside the classroom; and enabled the students to work with faculty to help make meaning of their construction and interpretation of knowledge.

Indeed, though research has examined the impact of peer support on academic success in both community colleges and four-year institutions (e.g., Tinto, 1997), to our knowledge, little research has explored this in relation to student success in STEM at community colleges. Nevertheless, research on students at four-year institutions has shown an inextricable connection between peer support and success for students in STEM. Evidence suggests that peers can facilitate minority student success by serving the important function of role modeling (Bandura, 1997; Museus et al., 2011). More specifically, when students observe peer role models succeed in STEM fields, they can recognize that they have the potential to succeed in STEM as well (Museus et al., 2011). Researchers have also underscored the importance of students and their peer role models sharing important social characteristics, such as race (Murphey, 1995; Museus et al., 2011), because those similarities enhance the message that the observer is capable of achieving the same task as the role model. Thus, minority students who are academically successful can send messages to their minority peers that the latter can succeed as well.

Conversely, empirical research on the impact of peers on minority students' success in STEM indicates that the type of interaction that students have with their peers does matter. Cole and Espinoza (2008), for example, conducted a longitudinal study of 146 Hispanic students in STEM and found that involvement in diversity functions and studying with other students was negatively associated with their GPA. These findings could be due to the fact that such involvement, if it is outside of the STEM environment, can function to marginalize minority students from the cultures of STEM disciplines (Bonous-Hammarth, 2000). In contrast, when minority students are involved in pre-professional clubs in STEM, they are more likely to persist (Hurtado et al., 2007; Museus et al., 2011). In addition, when studying in STEM-related peer groups, working with peers in science laboratories, and sharing information with upperclassmen, racial and ethnic minorities are more likely to succeed (Museus et al., 2011).

As demonstrated by the review of literature, there are various factors that influence the success of Black male community college students in STEM. To illustrate those factors as well as the relationships that exist among them, the subsequent section of this chapter will provide a model of "Success for Black Males in STEM." Finally, this chapter will conclude with implications for future research, policy, and practice for STEM at the community college level.

Several components comprise the Model of Success for Black Males in STEM. The first construct on the left indicates how Black males' experiences in K-12, shaped by systematic inequities, can restrict their academic preparedness as they enter community colleges. The subsequent constructs indicate that psychological factors (e.g., self-concept and self-efficacy); peer support (e.g., peer role models, STEM clubs, and learning communities); academic advising (e.g., quality of advising); supportive and affirming institutional climates; faculty support (e.g., culturally relevant pedagogy, research opportunities, and supportive relationships); and external factors (e.g., availability of financial aid; familial support; and STEM related employment opportunities) positively contribute to the success for Black male community college students in STEM. Conversely, the factors illustrated at the bottom of the model impair the success of Black students in STEM.

Each of the constructs illustrated and described above may be informative to researchers, policymakers, and practitioners who wish to understand the

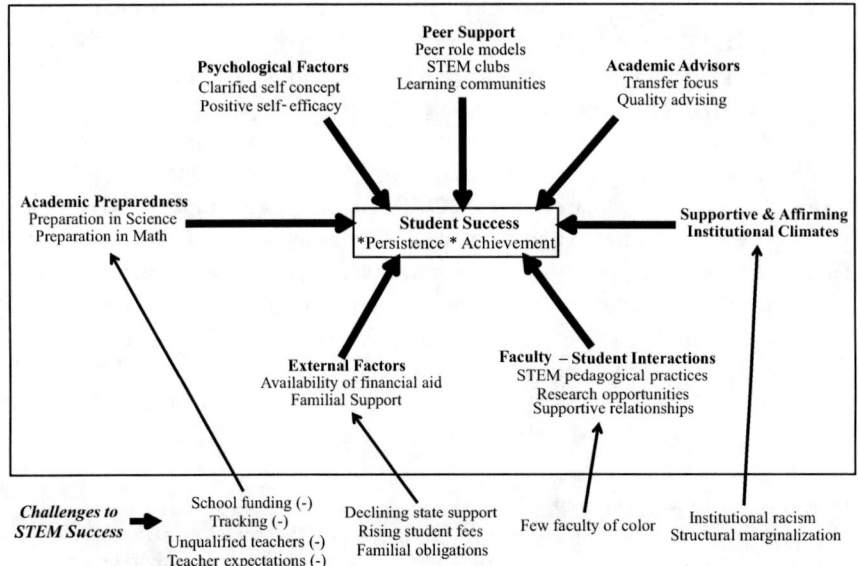

FIGURE 12.1 Palmer and Dubord's (2013) model of success for Black Men in STEM.

factors that influence Black male success in STEM in postsecondary education. The K-12 factors in the model are critical to the cultivation of strong academic preparedness among students in STEM education in community college. The college-level factors are important in efforts to increase persistence or facilitate the institutional transfer for Black male community college students in STEM. The relationships between these constructs are related to the implications for research, policy, and practice to which we now turn.

Implications for Future Research, Policy, and Practice

The preceding literature review and emergent model underscore several implications for future research, policy, and practice.

Future Research

A review of the existing literature suggests that additional research is needed to further the development of a model of success for Black male persistence in community colleges and STEM. Broadly speaking, research is needed on the experiences of Black males in community colleges in general. Specifically, this research ought to consider how Black males experience the institutional climates of community colleges, and how these climates influence the success of Black males in STEM.

Similarly, more information about the relationships that Black males form with faculty at community colleges is needed. In particular, research is needed on the pedagogical practices of faculty at community colleges, especially those teaching STEM courses. A greater understanding of what pedagogical factors support or undermine the success of Black males in STEM is necessary to further develop the model of success. The extent to which Black males are involved with faculty research in STEM at community colleges is also unclear. Although some community colleges are participating in the MESA Community College Program (Starobin et al., 2013), further research on this topic may provide insight into the extent to which Black males engage in research with faculty. In addition, faculty may, in some cases, serve as academic advisors. Given that the impact of academic advising on the success of students in STEM remains unclear (Museus et al., 2011), more research is needed in general on academic advising and its impact on the success of Black males in STEM.

Policy

Ensuring academic preparedness is critical to student success in STEM, and the research reviewed in this chapter highlights several policy implications to bolster the academic preparedness of Black males for postsecondary STEM education. Broadly speaking, action must be taken to improve the K-12 experiences of minority students.

A critical step to improving the educational experience of minority students in K-12 is ensuring that all students have access to instructors that are qualified and experienced. Indeed, Stotko, Ingram, and Beaty-O'Ferrall (2007) argued that a successful recruitment of qualified teachers to urban school districts should include, among other things, opportunities for applicants to interact with recruiters who are themselves successful urban teachers; information about mentoring programs for new teachers; profiles of student success stories; and information about the quality of life and professional opportunities from living and working in an urban environment. Urban districts should also streamline their hiring calendar, as applicants discouraged by long wait times will frequently give up and apply to suburban school districts that respond more quickly (Stotko et al., 2007). Indeed, research has also shown that giving K-12 teachers mentorship opportunities, control over curriculum, culturally engaging induction programs, and ongoing professional development contributes to retaining experienced and qualified teachers.

Beyond being qualified and experienced, K-12 teachers must also have high expectations of all students. Part of having high expectations of students is to eliminate or significantly reduce the prevalence of academic tracking and to provide equitable access to advanced educational opportunities. For instance, Lewis (2003) found that Black students at some inner-city schools do not have access to advanced placement courses; this should be rectified.

Finally, funding for both K-12 education and community colleges should be improved. Reducing the funding disparity between suburban and urban school districts creates more equitable academic opportunities for minority students. Similarly, increasing the funding for community colleges will allow those institutions to hire additional faculty and provide scholarships to students for whom cost is a barrier.

Practice

There are existing opportunities for community colleges to buoy the efforts of Black male students in STEM. Given the importance of academic preparation for success in STEM, partnerships between community colleges and K-12 schools can enhance preparedness for higher education. Such partnerships would be particularly meaningful with institutions that are otherwise unable to provide students with advanced placement courses or other more rigorous learning opportunities.

After students enter community college campuses, institutions should take action to create an environment with ample opportunity to establish positive relationships with peers and faculty. Community colleges can create spaces and programming that allow STEM students to interact and build relationships with their peers or other campus agents such as student support personnel and academic advisors. Community colleges can also provide workshops and training to students in STEM that will help them develop or enhance their self-efficacy and self-concept.

Aside from providing meeting space and offering programming, community colleges must ensure that all students have access to supportive academic advisors that provide quality guidance. It is especially important that community colleges provide superb academic advisors as they play a key role in assisting students seeking to transfer to four-year institutions.

In their teaching role, it is imperative that faculty present themselves as friendly and helpful; monitor student progress and express concern when necessary; listen to student concerns; and encourage students to strive for academic excellence. Further, faculty should seek to make STEM materials culturally relevant to the students in their classes. In essence, faculty should display an ethic of care when interacting with students, especially minority students. Community colleges should also encourage faculty in STEM to include students in research opportunities, as this has a number of benefits, including deeper understanding of STEM disciplines, more faculty–student interaction, among others.

References

Bahr, P. R. (2008). Cooling out in the community college: What is the effect of academic advising on students' chances of success? *Research in Higher Education, 49*(8), 704–732.

Bandura, A. (1997). *Self-efficacy: The exercise of control*. New York, NY: W. H. Freeman and Company.

Beckles, W. A. (2008). *Redefining the dream: African American male voices on academic success* (doctoral dissertation). Available from ProQuest Dissertations and Theses database. (UMI No. 3314150)

Bonous-Hammarth, M. (2000). Pathways to success: Affirming opportunities for science, mathematics, and engineering majors. *Journal of Negro Education, 69*(1-2), 92–111.

Bush, E. C. (2004). Dying on the vine: a look at African American student achievement in California community colleges. *Dissertation Abstracts International, 64*(12).

Bush, E. C., & Bush, L. (2005). Black male achievement and the community college. *Black Issues in Higher Education, 22*(2), 44.

Cabrera, A. F., Crissman, J. L., Bernal, E. M., Nora, A. P. T., & Pascarella, E. T. (2002). Collaborative learning: Its impact on college students' development and diversity. *Journal of College Student Development, 43*, 20–34.

Cole, D., & Espinoza, A. (2008). Examining the academic success of Latino students in science technology engineering and mathematics (STEM) majors. *Journal of College Student Development, 49*(4), 285–300.

DeSousa, D. J. (2005). *Promoting student success: What advisors can do* (Occasional Paper No. 11). Bloomington: Indiana University Center for Postsecondary Research.

Esters, L. L., & Mosby, D. C. (2007). Disappearing acts: The vanishing Black male on community college campuses. *Diverse Issues in Higher Education, 24*(14), 45.

Fenske, R. H., Porter, J. D., & DuBrock, C. P. (2000). Tracking financial aid and persistence of women, minority, and needy students in science, engineering, and mathematics. *Research in Higher Education, 41*(1), 67–94.

Flores, A. (2007). Examining disparities in mathematics education: Achievement gap or opportunity gap? *High School Journal, 91*(1), 29–42.

Flowers, L. A. (2006). Effects of attending a 2 year institution on African American males' academic and social integration in the first year of college. *Teachers College Record, 108*(2), 267–286.

Gonzalez, K. P. (2003). Campus culture and the experiences of Chicano students in a predominantly White university. *Urban Education, 37*(2), 193–218.

Grandy, J. (1998). Persistence in science of high-ability minority students. *Journal of Higher Education, 69*(6), 589–620.

Guiffrida, D. A, (2005). Othermothering as a framework for understanding African American students' definitions of student-centered faculty. *The Journal of Higher Education, 76*(6), 701–723.

Hagedorn, S. L., Cabrera, A., & Prather, G. (2010). The community college transfer calculator: Identifying the course-taking patterns that predict transfer. *Journal of College Student Retention, 12*(1), 105–130.

Hagedorn, S. L., Maxwell, W., & Hampton, P. (2001–2002). Correlates of retention for African-American males in the community college. *Journal of College Student Retention, 3*(3), 243–263.

Hagedorn, L. S. & Purnamasari, A. V. (2012). A realistic look at STEM and the role of the community colleges. *Community College Review, 40*(2), 145–164.

Harper, S. R., & Newman, C. B. (Eds.). (2010). *Students of color in STEM* (New Directions for Institutional Research, No. 148). San Francisco, CA: Jossey-Bass.

Hurtado, S., Han, J. C., Sáenz, V. B., Espinosa, L. L., Cabrera, N., & Cerna, O. S. (2007). Predicting transition and adjustment to college: Biomedical and behavioral science aspirants' and minority students' first year of college. *Research in Higher Education, 48*(7), 841–887.

Jordan, P. G. (2008). *African American male students' success in an urban community college: A case study* (doctoral dissertation). University of Pennsylvania, Philadelphia.

King, M. C. (1993). *Academic advising: Organizing and delivering services for student success* (New Directions for Community Colleges No. 82). San Francisco, CA: Jossey-Bass.

Leppel, K. (2002). Similarities and differences in the college persistence of men and women. *Review of Higher Education: Journal of the Association for the Study of Higher Education, 25*(4), 433–450.

Lewis, B. F. (2003). A critique of literature on the under-representation of African Americans in science: Directions for future research. *Journal of Women and Minorities in Science and Engineering, 9*(3&4), 361–373.

Malcom, L. E. (2010). Charting the pathways to STEM for Latina/o students: The role of community colleges. In S. R. Harper & C. B. Newman (Eds.), *Students of color in STEM: Engineering a new research agenda: New directions for institutional research* (pp. 29–40). San Francisco, CA: Jossey-Bass.

Maton, K. I., & Hrabowski, F. A. (2004). Increasing the number of African American PhDs in the sciences and engineering. *American Psychologist, 59*(6), 547–556.

Maton, K. I., Hrabowski, F. A., & Schmitt, C. L. (2000). African American college students excelling in the sciences: College and postcollege outcome in the Meyerhoff Scholars Program. *Journal of Research in Science Teaching, 37*(7), 629–654.

McArthur, R. C. (2005). Faculty-based advising: An important factor in community college retention. *Community College Review, 32*(4), 1–19.

Murphey, T. (1995). Identity and beliefs in language learning. *The Language Teacher, 19*(4), 34–36.

Museus, S. D., Palmer, R. T., Davis, R. J., & Maramba, D. C. (2011). Racial and ethnic minority student's success in STEM education. *ASHE-Higher Education Report Series, 36*(6), 1-140.

National Science Foundation. (2006). *Science, technology, engineering, and mathematics talent expansion program.* Retrieved from http://www.nsf.gov.funding/pgm_summ.jsp?pims_id=5488

National Science Foundation. (2008). *Science and engineering indicators 2008* (Report No. NSB-08-01A). Arlington, VA: Author.

National Science Foundation. (2010). *Classification of programs.* Washington, DC: Author. Retrieved from http://www.nsf.gov/statistics/nsf99330/pdf/sectd.pdf.

Nevarez, C., & Wood, J. L. (2010). *Community college leadership and administration: Theory, practice, and change.* New York, NY: Peter Lang.

Palmer, R. T., & Maramba, D. C. (2012). Creating conditions of mattering to enhance persistence for Black males at an historically Black University. *Spectrum: Journal on Black Men, 1*(1), 95–120.

Palmer, R. T., Maramba, D. C., & Dancy, T. E. (2011). A qualitative investigation of factors promoting the retention and persistence of students of color in STEM. *Journal of Negro Education, 80*(4), 491-504.

Pascarella, E. T., & Terenzini, P. T. (2005). *How college affects students: A third decade of research.* San Francisco, CA: Jossey-Bass.

Paulsen, M., & St. John, E. (2002). Social class and college costs: Examining the financial nexus between college choice and persistence. *Journal of Higher Education, 73*(2), 189–236.

Perin, D. (2005). Institutional decision making for increasing academic preparedness in community colleges. *New Directions for Community Colleges, 129,* 27–38.

Perna, L., Lundy-Wagner, V., Drezner, N. D., Gasman, M., Yoon, S., Bose, E., & Gary. S. (2009). The contribution of HBCUs to the preparation of African American women for STEM careers: A case study. *Research Higher Education, 50*(1), 1–23.

Rhoades, G. (2012). Closing the door, increasing the gap. Who's not going to community college? Retrieved from http://www.futureofhighered.org

Seymour, E., & Hewitt, N, M. (1997). *Talking about Leaving: Why undergraduates leave the sciences.* Oxford, England: Westview Press.

Starobin, S. S., Laanan, F. S., & Burger, C. J. (2010). Role of community colleges: Broadening participating among women and minorities in STEM. *Journal of Women and Minorities in Science and Engineering, 16*(1), 1–5.

Starobin, Jackson, D., & Laanan, F. S. (2013). Model programs for STEM student success at minority serving two-year colleges In R. T. Palmer, D. C. Maramba, & M. Gasman (Eds.), *Fostering success of ethnic and racial minorities in STEM: The role of minority serving institutions* (pp. 59–71). New York, NY: Routledge.

Stotko, E. M., Ingram, R., & Beaty-O'Ferrall, M. E. (2007). Promising strategies for attracting and retaining successful urban teachers. *Urban Education, 42,* 30–51.

Strayhorn, T. L. (2010). When race and gender collide: Social and cultural capital's influence on the academic achievement of African American and Latino males. *Review of Higher Education, 33*(3), 307-332.

Tinto, V. (1993). *Leaving college: Rethinking the causes and cures of student attrition (Vol. 2).* Chicago, IL: University of Chicago Press.

Tinto, V. (1997). Classroom as communities: Exploring the educational character of student persistence. *Journal of Higher Education, 68*(5), 659–673.

Wood, J. L. (2011). Leaving the 2 year college: Predictors of Black male collegians departure. *Journal of Black Studies, 43*(3), 303–326.

Wood, J. L., & Turner, C. S. V. (2011). Black males and the community college: Student perspectives on faculty and academic success. *Community College Journal of Research and Practice, 35,* 1–17.

ABOUT THE EDITORS

Robert T. Palmer, PhD, is an Assistant Professor of Student Affairs Administration at the State University of New York, Binghamton. Since completing his PhD in 2007, Dr. Palmer's work has been published in national referred journals, and he has authored well over 75 refereed journal articles, book chapters, and other academic publications. His books include *Racial and Ethnic Minority Students' Success in STEM Education* (2011, with Samuel Museus, Ryan J. Davis, and Dina C. Maramba), *Black Men in College: Implications for HBCUs and Beyond* (2012, with J. Luke Wood), *Black Graduate Education at HBCUs: Trends, Experiences, and Outcomes* (2012, with Adriel A. Hilton and Tiffany Fountaine), *Fostering Success of Ethnic and Racial Minorities in STEM: The Role of Minority Serving Institution* (2012, with Dina C. Maramba and Marybeth Gasman), and *STEM Models of Success: Programs, Policies, and Practices* (forthcoming, with J. Luke Wood). In 2009, the American College Personnel Association's (ACPA) Standing Committee for Men recognized his excellent research on Black men with its Outstanding Research Award. In 2011, Dr. Palmer was named an ACPA Emerging Scholar and, in 2012, he received the Carlos J. Vallejo Award of Emerging Scholarship by the Multicultural/Multiethnic Education SIG of the American Education Research Association. Furthermore in 2012, he was awarded the ASHE-Mildred García Junior Exemplary Scholarship Award.

J. Luke Wood, PhD, is an Assistant Professor of Administration, Rehabilitation, and Postsecondary Education at San Diego State University (SDSU). Dr. Wood is Co-Director of the Minority Male Community College Collaborative (M2C3), Chair of the Multicultural & Multiethnic Education (MME) special interest group of the American Educational Research Association (AERA), and Chair-Elect for the Council on Ethnic Participation (CEP) for the Association for the Study of Higher Education (ASHE). Wood's research focuses on factors impacting the success of Black (and other minority) male students in the community college. In particular, his research examines contributors (e.g., social, psychological, academic, environmental, institutional) to positive outcomes (e.g., persistence, achievement, attainment, transfer, labor market outcomes) for these men. Wood is the co-author of two textbooks, *Community College Leadership & Administration: Theory, Practice, and Change* (2010, with Carlos Nevarez), and *Leadership Theory in the Community College: Applying Theory to Practice* (2013, with Carlos Nevarez and Rose Penrose). He is also co-editor of the books, *Black Men in College: Implications for HBCUs and Beyond* (2012, with Robert T. Palmer), and *Black Males in Postsecondary Education: Examining Their Experiences in Diverse Institutional Contexts* (2012, with Adriel Hilton and Chance Lewis).

ABOUT THE CONTRIBUTORS

Lori Andersen is a doctoral student in Educational Policy, Programming, and Leadership at the College of William and Mary, specializing in Gifted Education. Ms. Andersen is a National Board Certified Teacher in Adolescent and Young Adult Science who has taught physics at the community college and high school levels.

James E. Bartlett, II, is an associate professor and director of the adult and community college education doctoral cohort program at North Carolina State University. Previously Dr. Bartlett has had faculty appointments at the University of South Carolina and the University of Illinois Urbana-Champaign. His research has been funded by the U.S. Department of Education, National Center for Career and Technical Education, as well as other public and private organizations. Dr. Bartlett has consulted in a variety of agencies in a number of sectors including telecommunications, military, human resource management, and service industries. Dr. Bartlett's work has been published in a number of national refereed journals and he has well over 100 refereed journal articles, book chapters, and other academic publications. His work has appeared in publications such as *Advances in Human Resource Development, Journal of Vocational Education Research, Delta Pi Epsilon Journal,* and *Behavior Research Methods, Instruments & Computers.* Much of Dr. Bartlett's work seeks to improve research by the development of faculty. Dr. Bartlett has served as the Chair the American Education Research Association Workplace Learning Special Interest Group and President of the Association for Career and Technical Education Research. Dr. Bartlett has also served as the co-chair the quantitative methods special interest group for the Academy of Human Resource Development. Recently James received the 2012 ACTER outstanding symposium award for a session

"Developing Faculty to Improve Research Methods for CTE Research." Dr. Bartlett has served on the editorial boards of the *Journal of Career and Technical Education* and *Career and Technical Education Research* as well as editor of both journals.

Estela Mara Bensimon, PhD, is a Professor of Higher Education and co-director of the Center for Urban Education (CUE) at the USC Rossier School of Education. Her current research is on issues of racial equity in higher education from the perspective of organizational learning and socio-cultural practice theories. Dr. Bensimon's publications about equity, organizational learning, practitioner inquiry and change include: *The Underestimated Significance of Practitioner Knowledge in the Scholarship on Student Success; Doing Research that Makes a Difference; Equality in Fact, Equality in Results: A matter of institutional accountability; Measuring the State of Equity in Public Higher Education* and *Closing the Achievement Gap in Higher Education: An Organizational Learning Perspective.*

Megan M. Chase is a research assistant at the Center for Urban Education (CUE) and doctoral candidate in the Urban Education program at the USC Rossier School of Education. Megan is currently working with her advisor, Dr. Estela Mara Bensimon, examining issues of equity in state level higher education policy.

Katherine Mary Conway, PhD, is deputy chair of the Business Management Department at Borough of Manhattan Community College, City University of New York (CUNY). Conway's research is focused on community college student access and persistence, with an emphasis on immigrant, minority, and first-generation students. Conway has presented her work at the Association for the Study of Higher Education, American Educational Research Association and the American Association of University Professors.

Marjorie L. Dorimé-Williams is a doctoral candidate in Educational Policy and Organizational Leadership at the University of Illinois at Urbana-Champaign. Her research interests include persistence and retention of Black students in post-secondary education, student support programming in college, identity intersectionality with a focus on race and class, and program evaluation.

Alicia C. Dowd, PhD, is Associate Professor of higher education at the University of Southern California's Rossier School of Education and co-director of the Center for Urban Education (CUE). Dr. Dowd's research focuses on political-economic issues of racial-ethnic equity in postsecondary outcomes, organizational learning and effectiveness, accountability and the factors affecting student attainment in higher education. Dr. Dowd has served as the principal investigator of several major, national studies of institutional effectiveness, equity, community college transfer, benchmarking, and assessment.

About the Contributors

Zachary M. DuBord is a graduate student at the State University of New York, Binghamton, where he studies Student Affairs Administration. Prior to being a student, Zach worked in student disability services and as library branch manager.

Pamela Eddy, PhD, is an Associate Professor in Educational Policy, Planning, and Leadership at the College of William and Mary. Her research interests include community college leadership and development, organizational change and educational partnerships, gender roles in higher education, and faculty development. Eddy was the President of the Council for the Study of Community Colleges 2011-2012 and currently serves as past-president of the council. She received the 2006 emerging scholar award by the Council for the Study of Community Colleges, the 2007 Central Michigan University Provost Award for Research and Creative Endeavors, the 2008 Central Michigan University Teaching Excellence Award Winner, the 2011 Plumeri Award for Faculty Excellence at the College of William and Mary, and was a Fulbright Scholar in Dublin, Ireland.

Edward C. Fletcher Jr., PhD, is Assistant Professor in the Department of Adult, Career and Higher Education at the University of South Florida. He is the coordinator of the Master in Career and Technical Education. His primary research interest focuses on longitudinally studying the influence of high school curriculum tracking on labor market and postsecondary outcomes in adulthood.

Bobbie Everett Frye is the Director of Institutional Research at Central Piedmont Community College. Her educational background is in sociology and education and she currently serves several advisory and leadership roles for student success initiatives. She has co-authored several studies at CPCC including the Millennial, Minority Male Mentoring research and is currently researching developmental math students. She has presented her research at several conferences including the Association of Institutional Research (AIR), Council for the Study of Community Colleges (CSCC) and the National Center for Community College Student Engagement (NCCCSE). Her research interests include developmental education, low income and underrepresented community college students. She has an extensive background in research design, statistical methods, software and programming knowledge in (SPSS and SAS). She is pursuing a doctorate in education at North Carolina State University.

Victor Hernandez-Gantes, PhD, is Associate Professor in the Department of Adult, Career and Higher Education at the University of South Florida. He is the director of the Career and Workforce Education program. His research interests focus on the interface that integrates the design, implementation, and evaluation of educational strategies designed to connect curriculum, teaching, and learning in work contexts as a means to maximize career-oriented learning for all students.

Ginelle John, PhD, is the Enrollment Administrator for the Department of Occupational Therapy at New York University. In 2011 Ginelle earned her PhD in Higher Education from New York University. Her research interests include college access and retention and college athletics. Ginelle's recent publications focused on underrepresented student majoring and persisting in STEM majors and on community college athletes. In 2010 Ginelle was awarded a NCAA Graduate Student Research Grant for her project Black Male College Athletes: Capital and Educational Outcomes. Her study examined the relationship between Black college football and basketball players' SES, high school social and academic activities, college activities, and their educational outcomes.

Royel M. Johnson is a doctoral student within the Department of Education Policy, Organization & Leadership at the University of Illinois at Urbana-Champaign. Currently, he serves as the Vice-President for Administration for the National Black Graduate Student Association. His research interests center on two major streams of scholarly inquiry: (a) student access and achievement, and (b) the study of inequality and diversity in education. Much of his research attention is devoted to the experiences of historically underrepresented groups in education.

Tiffany Jones is a research assistant at the Center for Urban Education (CUE). Tiffany works with Estela Mara Bensimon on higher education, critical policy analysis, and implementing the equity scorecard at institutions populated by predominately students of color.

Valerie C. Lundy-Wagner, PhD, is an Assistant Professor and Faculty Fellow at New York University in the Higher & Postsecondary Education Program at the Steinhardt School for Culture, Education, and Human Development. Her research is primarily concerned with bachelor's degree completion, and what colleges and universities can to do better conceptualize and enact programs and policies that help students persist. Lundy-Wagner's work also includes theoretical, historical, and quantitative research often related to the postsecondary science, technology, engineering, and mathematics (STEM) pipeline or historically Black colleges and universities (HBCUs).

Dina C. Maramba, PhD, is Associate Professor of Student Affairs Administration and affiliate faculty with Asian and Asian American Studies at the State University of New York, Binghamton. With over 10 years of experience as a student affairs practitioner, she worked in many roles including working closely with STEM students and facilitating their success in college. She is also one of the co-authors of an ASHE-Higher Education Report about racial and ethnic minority students in the STEM educational pipeline. Dr. Maramba is a recipient of the 2011 Award for Outstanding Contribution to Asian/Pacific Islander American Research Relating to Higher Education by the Association of College Personnel Administrators (ACPA).

Cecilia Santiago is a Project Specialist at Center for Urban Education (CUE) at the University of Southern California. Her dissertation work focuses on identifying the ways in which faculty act as institutional agents for low-income Latino students in STEM disciplines at HSIs.

Linda Taing Shieh is a Research Assistant working with Dr. Alicia C. Dowd at the Center for Urban Education (CUE). Her most recent research with CUE looks at community college student access and success at liberal arts institutions using the equity scorecard framework. Her research interests include examining how policies and social structures impact access for racial ethnic minorities in higher education, and the role of practitioners in the success of student outcomes.

Kelly D. Smith earned a BA in Chemistry from Case Western Reserve University, a MS in Chemistry from the University of Illinois at Urbana-Champaign, and a MS in Forensic Science from The George Washington University. After working for more than nine years in forensic science, Kelly moved into the field of education. She is starting her ninth year as a secondary science instructor at Sun Valley High School in Union County, North Carolina, and is currently working on a doctorate in education at North Carolina State University, where her research interests include student success and instructor job satisfaction.

Frances King Stage, PhD, is Professor of Administration, Leadership, and Technology at New York University. Her research specialization includes college student learning, especially for STEM disciplines and student participation in math and science majors. Recent work has focused on characteristics of undergraduate institutions that produce unexpected levels of students who go on to earn STEM doctorates. Stage is past Vice President for the Postsecondary Education Division of the American Educational Research Association and has won awards for research and scholarship from the Association for the Study of Higher Education and the American Educational Research Association. She spent 1999-2000 as a Senior Fellow at the National Science Foundation and was a Fulbright Specialist at the University of West Indies, Mona, Jamaica in 2008 and at the University of West Indies, Cave Hill, Barbados in 2011.

Terrell L. Strayhorn, PhD, is Associate Professor of Higher Education at The Ohio State University, where he also serves as Director of the Center for Higher Education Research and Policy (CHERP), Senior Research Associate in the Kirwan Institute for the Study of Race & Ethnicity, and Faculty Associate in the Todd Bell National Resource Center for Black Males. He holds joint appointments in the Department of African American & African Studies, Sociology, and Engineering Education. Professor Strayhorn maintains an active and highly visible research agenda focusing on major policy issues in education such as equity/diversity, access/retention, and student development, learning, and success. In 2011, Diverse Issues in Higher Education named him one of the nation's Top Emerging Scholars. Strayhorn is co-editor of Spectrum: A Journal on Black Men, published by Indiana University Press; associate editor of the NASAP Journal, and serves on several editorial boards. Dr. Strayhorn is a member of Alpha Phi Alpha Fraternity, Incorporated.

Derrick L. Tillman-Kelly is a doctoral student in the higher education and student affairs program at The Ohio State University. He also serves as a graduate research associate to Dr. Terrell Strayhorn and editorial assistant for Spectrum: A Journal on Black Men. His primary research interest focuses on identity intersection, specifically considering race, gender, sexuality and spirituality as social identities. Secondary interests include the organizational socialization of administrators in higher education.

Marissa Vasquez Urias is a doctoral student in the departments of Administration, Rehabilitation and Postsecondary Education and Educational Leadership at San Diego State University. Her research examines factors facilitating the success of Latino male students in community colleges.

Xueli Wang, PhD, is an Assistant Professor in the Department of Educational Leadership and Policy Analysis at the University of Wisconsin-Madison and a scholar at the Wisconsin Center for the Advancement of Postsecondary Education (WISCAPE). Wang's research centers on the secondary-postsecondary nexus and the intersection between motivational beliefs, social disadvantage, and college experience and success, with a particular focus on students beginning at community colleges. Wang's current research deals with several issues pertaining to community colleges and their students, including baccalaureate aspirations, participation in STEM fields of study, educational pathways, as well as the role of community colleges in STEM education.. In summer 2011, Wang was selected as a Young Academic Fellow by the Institute of Higher Education Policy and the Lumina Foundation.

Thomas J. Ward, PhD, is Professor and Associate Dean in the School of Education at the College of William and Mary in Williamsburg, Virginia. Among his primary research interests are the use of data modeling for teaching and school improvement, the use of test data in decision making, and at-risk programs evaluation. He has worked as a consultant with the state departments of education in Virginia, South Carolina, and Pennsylvania and with numerous school divisions in Virginia, Pennsylvania, New Jersey, and Delaware. His doctorate in educational psychology was received from the Pennsylvania State University.

Michael Steven Williams is a doctoral student in Higher Education and Student Affairs at The Ohio State University. Michael's research interests center on two aspects of higher education: (1) *the student*, particularly graduate student socialization and mentoring and (2) *the institution,* with focus on specialized institutions such as historically Black colleges and universities (HBCUs). Michael is also a member of Alpha Phi Alpha Fraternity, Inc.

Denise Yull, EdD, is an Assistant Professor of Human Development at the State University of New York, Binghamton. She comes to SUNY Binghamton after spending 15 years in a career as a Mechanical Engineer and 12 years as a Community College mathematics instructor. Following her passion to work in the field of education, Denise completed a MA in Mathematics and an EdD in Educational Theory and Practice both from SUNY Binghamton. Dr. Yull's work focuses on the influence of structural factors that impact educational disparities in marginalized communities in the context of secondary and higher education.

INDEX

Page numbers in italic refer to figures or tables

A
Action control, 81, 85
   defined, 77
Advisors
   Black males, 201
   math, 132
   student meetings, 83
Agency, concept, 174
American Graduation Initiative, 17
Articulation agreements, 29
   Asian American students, 168
   Pacific Islander students, 168
Asian American and Native American Pacific Islander Serving Institutions
   characterized, 163–164
   City College of San Francisco, 164–165
Asian American students
   articulation agreements, 168
   community colleges
      current initiatives, 163–166
      participation, 159–161
   disaggregated data, 166–167
   ethnic groups, 157–158, 166
   higher education, participation, 158–163
   remediation, 163
   STEM
      current initiatives, 163–166
      participation, 161–163
   student affairs, 167–168

Asian Pacific Islander American Scholarship Fund, 163, 165–166
Associate's degrees
   community colleges, 141–142
   minorities, 141–152
      data, 145–146
      results, 146–150, *147–148*
      study method, 145–146
   STEM, 23, 45, 141–152
      data, 145–146
      results, 146–150, *147–148*
      study method, 145–146

B
Beginning Postsecondary Students Longitudinal Study, 4, 8, *9,* 10, *10,* 12, *12*
Black male students
   community colleges, 193–206
      academic advisors, 201
      academic preparedness, 195–197
      characterized, 193–194
      external commitments, 198–200
      faculty support, 198–200
      faculty-student engagement in research opportunities, 200–201
      financial support, 197–198
      future research, 204
      institutional climate, 198–200
      Palmer and Dubord's model of success, *203*
      pedagogical approaches, 198–200
      peer support, 201–204

policy implications, 204–205
practice implications, 205–206
psychological indicators, 198–200
higher education, characterized, 193–194
STEM, 193–206
  academic advisors, 201
  academic preparedness, 195–197
  external commitments, 198–200
  faculty support, 198–200
  faculty-student engagement in research opportunities, 200–201
  financial support, 197–198
  future research, 204
  institutional climate, 198–200
  Palmer and Dubord's model of success, 203
  pedagogical approaches, 198–200
  peer support, 201–204
  policy implications, 204–205
  practice implications, 205–206
  psychological indicators, 198–200
Black students, *see also* Black male students
  STEM persistence, 61–72
    domain-specific self-efficacy, 67–68
    high-ability criteria, 64–65, *65*, 67
    model, 65, 68
    results, 65–68, *66*, *67*
    study method, 64–65
    subjective task value, 69–70
  subjective task value, 62

C

Career academies, 45, 47–48
Career options, community colleges, 70
Certificate, STEM, 45
Change
  STEM pipeline, 48–49
  Top Jobs Act, 27–28, 31–32, 33
Change theory, 20–21
City College of San Francisco, Asian American and Native American Pacific Islander Serving Institutions, 164–165
Collaborative learning
  characterized, 135
  math education, 133
  results, 135
College readiness, 29
Color-blindness, 178–179
Community College Student Experiences Questionnaire, 93, 95, 98–99
Community colleges
  Asian American students
    current initiatives, 163–166

participation, 159–161
associate's degrees, 92, 141–142
Black males, 193–206
  academic advisors, 201
  academic preparedness, 195–197
  characterized, 193–194
  external commitments, 198–200
  faculty support, 198–200
  faculty-student engagement in research opportunities, 200–201
  financial support, 197–198
  future research, 204
  institutional climate, 198–200
  Palmer and Dubord's model of success, 203
  pedagogical approaches, 198–200
  peer support, 201–204
  policy implications, 204–205
  practice implications, 205–206
  psychological indicators, 198–200
career options, 70
certificates, 92
cost, *93,* 93–94
credentialing function, 92
developmental courses, 9–10
ethnicity
  academic choices, 4–8, *6*
  achievement gap, 97
  adjustment and satisfaction, 94
  analysis, 95–97
  as entry point to postsecondary education, 3
  Beginning Postsecondary Students Longitudinal Study, 4, 8, *9,* 10, *10,* 12, *12*
  bridging research and policy, 13–14
  Education Longitudinal Study, 4, 5, *6*
  family and career responsibilities, 94
  foundational coursework, 11–12, *12*
  four-year STEM majors attending community colleges, 11–12, *12*
  initial choice of STEM major, 8–10, *9*
  intervention, 13
  lack of evidence-based knowledge, 13–14
  learning gains, 96–97, 98
  Postsecondary Education Transcript Study, 8
  previous research results, 93–95
  psychological factors, 91–99
  recent high school graduates, 4–8, *6*
  satisfaction, 96, 97
  STEM transfer, *10,* 10–11
  STEM-related aspirations, 4–8, *6*

Community colleges (*continued*)
  study data source, 95
  study sample, 95, *96*
  toward bachelor's degree, 11–12, *12*
  transfer function, 94
functions, 92
Hispanic students, 172–191
Latinos/as, 172–191
  agency study barriers to agency, 176–180
  agency study discussion, 183–189, *184–187*
  agency study findings, 175–176
  agency study methodology, 174–175, *175, 176*
  faculty, administrator or staff agency, 173–191
  out-of-class experiences, 173
  special programs, 172–173
  student agency stance, 173
  what motivates faculty, administrative leaders, and staff, 174
minorities, increased quality and quantity, 104–105
minority-serving institutions characterized, 142–144
Pacific Islander students
  current initiatives, 163–166
  participation, 159–161
race
  academic choices, 4–8, *6*
  achievement gap, 97
  adjustment and satisfaction, 94
  analysis, 95–97
  as entry point to postsecondary education, 3
  Beginning Postsecondary Students Longitudinal Study, 4, 8, *9,* 10, *10,* 12, *12*
  bridging research and policy, 13–14
  Education Longitudinal Study, 4, 5, *6*
  family and career responsibilities, 94
  foundational coursework, 11–12, *12*
  four-year STEM majors attending community colleges, 11–12, *12*
  initial choice of STEM major, 8–10, *9*
  intervention, 13
  lack of evidence-based knowledge, 13–14
  learning gains, 96–97, 98
  Postsecondary Education Transcript Study, 8
  previous research results, 93–95
  psychological factors, 91–99
  recent high school graduates, 4–8, *6*

  satisfaction, 96, 97
  STEM transfer, *10,* 10–11
  STEM-related aspirations, 4–8, *6*
  study data source, 95
  toward bachelor's degree, 11–12, *12*
  transfer function, 94
remedial education, 9–10, 102
STEM, 3–14, 124–125
  advantages, 22
  change theory, 20–21
  coordination, 28–30
  degree completion, 23
  educators nurturing, 92
  initiatives in community colleges, 144–145
  project background, 24
  recent research, 21–33
  state policy impact, 17–33
  state policy implementation, 19–20
  strategic planning, 20–21
  study analysis, 30–33
  study findings, 24–30
  Virginia, 17–33
  Virginia case background, 22–23
students of color
  action control, 77, 81, 85
  control variables, 81–82
  degree utility, 77, 81, 85
  disparities, 75–76
  English self-efficacy, 81
  faculty-student interactions, 82–83
  field changes to non-STEM majors, 76, *76*
  internet use to access library resources, 84–85
  locus of control, 77, 81, 85
  math self-efficacy, 80–81
  meeting with advisors about academic plans, 83
  non-cognitive predictors, 76–87
  non-cognitive predictors implications, 79, 85–87
  non-cognitive predictors importance, 77–78
  research and practice recommendations, 86–87
  self-efficacy, 77, 85
  STEM, 75–76
  study analytic procedure, 82
  study findings, 82–85
  study methods, 79–80
  study variables, 80–82
  studying at library, 83–84

technician preparation, 45
Virginia
  degree completion, 23
  STEM, 17–33
  vocational job, 92
Competitiveness, 37
  innovation, 43
  STEM, 37
Computer science, 40, 41
Coordination, STEM, 19
Cost, remedial education, 101
Cultural competence, math education, 131–133
Culturally relevant teaching practices
  math education, 133–134, 135
    community colleges, 134
    propositions, 134–135
Culture, subjective task value, 63

D

Degree utility, 81, 85
  defined, 77
Demand/supply model, 46
Developmental courses
  community colleges, 9–10
  STEM, 9–10
Developmental math
  avoidance vs. completing, 103–104
  propensity score matching
    analyses, 113–115, *119–120*
    assessment, 106
    conceptual framework, 110–111
    descriptive results, 113, *119–120*
    environmental factors, 109–110
    independent variables, 110–111
    lack of consistency, 106
    methodology, 111–115
    placement, 106
    pre-screening, 112–113
    pre-treatment covariates selection, 112–113
    predictors, 116
    study population, 111–112
    study purpose, 103
    theoretical framework, 109–110
Direct instruction approach, math education, 129
Diversity
  higher education, 93
  STEM, 3–14
  STEM persistence, 59–72
Domain-specific self-efficacy, 61
Drill and practice pedagogy, math education, 129–130

Dual-enrollment programs, 29–30

E

Economic development, STEM, 19
Education Longitudinal Study, 4, 5, *6,* 79–80
  described, 5
Education, *vs.* workforce development, 38
Engineering, *See* STEM
English self-efficacy, 81
Ethnicity
  community colleges
    academic choices, 4–8, *6*
    achievement gap, 97
    adjustment and satisfaction, 94
    analysis, 95–97
    as entry point to postsecondary education, 3
    Beginning Postsecondary Students Longitudinal Study, 4, 8, *9,* 10, *10,* 12, *12*
    bridging research and policy, 13–14
    Education Longitudinal Study, 4, 5, *6*
    family and career responsibilities, 94
    foundational coursework, 11–12, *12*
    four-year STEM majors attending community colleges, 11–12, *12*
    initial choice of STEM major, 8–10, *9*
    intervention, 13
    lack of evidence-based knowledge, 13–14
    learning gains, 96–97, 98
    Postsecondary Education Transcript Study, 8
    previous research results, 93–95
    psychological factors, 91–99
    recent high school graduates, 4–8, *6*
    satisfaction, 96, 97
    STEM transfer, *10,* 10–11
    STEM-related aspirations, 4–8, *6*
    study data source, 95
    study sample, 95, *96*
    toward bachelor's degree, 11–12, *12*
    transfer function, 94
  STEM, 3–14, 156
    academic choices, 4–8, *6*
    achievement gap, 97
    adjustment and satisfaction, 94
    Beginning Postsecondary Students Longitudinal Study, 4, 8, *9,* 10, *10,* 12, *12*
    bridging research and policy, 13–14
    Education Longitudinal Study, 4, 5, *6*
    family and career responsibilities, 94
    foundational coursework, 11–12, *12*

Ethnicity (*continued*)
    four-year STEM majors attending community colleges, 11–12, *12*
    intervention, 13
    lack of evidence-based knowledge, 13–14
    learning gains, 96–97, 98
    participation trends, 38–42
    Postsecondary Education Transcript Study, 8
    previous research results, 93–95
    psychological factors, 91–99
    recent high school graduates, 4–8, *6*
    satisfaction, 96, 97
    STEM transfer, *10*, 10–11
    STEM-related aspirations, 4–8, *6*
    study data source, 95
    study sample, 95, *96*
    toward bachelor's degree, 11–12, *12*
    transfer function, 94
    underrepresentation, 38
    subjective task value, 63
Ethnomathematics
    advantages, 133–134
    characterized, 133–134
Expectancy-value model
    characterized, 60
    STEM persistence of ninth-grade, underrepresented minority students, 60–72
Expectancy-value theory, subjective task value, 62–63

F
Faculty advisor, math, 132
Faculty-student interactions, 82–83
    gender, 82–83
    race, 82–83
Family-oriented culture, 179–180

G
Gender, *See also* Black males; Women
    faculty-student interactions, 82–83
    interest usage, 84–85
    library use, 83–84
    role models, 123
    STEM
        participation trends, 38–42
        underrepresentation, 38
    STEM persistence, 65, *66*, 67
        domain-specific self-efficacy, 67–68
        high-ability criteria, 64–65, *65,* 67
        model, 65, 68
        results, 65–68, *66,* 67
        study method, 64–65
        subjective task value, 69–70
Geographical distribution
    innovation, 44–45
    research, 44–45
    STEM, 41
Global Competitiveness Report, 37

H
High school
    career academies, 45, 47–48
    math, 7
    ninth-grade minority students expectancy-value model, 60–72
    science, 7
    teaching capacity, 44, 45
    technician preparation, 45
High School Longitudinal Study, 64
Higher education, *see also* Specific type
    Asian American students, participation, 158–163
    Black males, characterized, 193–194
    cost, *93,* 93–94
    diversity, 93
    Pacific Islander students, participation, 158–163
Hispanic students, *see also* Latino/a students
    community colleges, 172–191
    STEM persistence, 61–72
        domain-specific self-efficacy, 67–68
        high-ability criteria, 64–65, *65,* 67
        model, 65, 68
        study method, 64–65
        subjective task value, 69–70

I
Implementation research, 19–20
Implementation, mobilization for, 32
Innovation
    competitiveness, 43
    cultural barrier, 29
    diffusion incentives, 44
    geographical distribution, 44–45
    National Science Foundation, 43
    STEM, promoting, 43–44
Institutional agent, 174
Institutional saga, 30–31
Integrated Postsecondary Education Data System, 145
Integrated workforce development systems
    minorities, underrepresented students, 38, 40, 51–52
    STEM

need for, 46–50
underrepresented students, 37–52
women, underrepresented students, 38, 40, 51–52
Interest usage
gender, 84–85
race, 84–85
inth-grade students, minorities
expectancy-value model, 60–72
Introductory science courses, 70–71

J
Journal of Women and Minorities in Science and Engineering, 75

L
Latino/a students, *See also* Ethnicity; Hispanic students; Minorities
community colleges, 172–191
agency study barriers to agency, 176–180
agency study discussion, 183–189, *184–187*
agency study findings, 175–176
agency study methodology, 174–175, *175, 176*
faculty, administrator or staff agency, 173–191
out-of-class experiences, 173
special programs, 172–173
student agency stance, 173
what motivates faculty, administrative leaders, and staff, 174
practitioner agency
actions reflecting agency, 180–183
assisting Latinos/as to become scientists, 181–182
barriers to agency, 176–180
color-blindness, 178–179
data practices, 177–178
developing college-knowledge in the Latino community, 182
family-oriented culture, 179–180
findings, 175–176
methodology, 174–175, *175, 176*
one-sided transfer articulation, 177
program building, 181–183
self-directed remediation of teaching practices, 182–183
structural barriers, 177–178
values and beliefs, 178–180
STEM, 172–191
agency study barriers to agency, 176–180
agency study discussion, 183–189, *184–187*
agency study findings, 175–176
agency study methodology, 174–175, *175, 176*
faculty, administrator or staff agency, 173–191
out-of-class experiences, 173
special programs, 172–173
student agency stance, 173
what motivates faculty, administrative leaders, and staff, 174
Library use
gender, 83–84
race, 83–84
Locus of control, 81, 85
defined, 77
Louis Stokes Alliances for Minority Participation (LSAMP) Program, 104–105, 144

M
Math anxiety, 127–128
Math education, *See also* STEM
as STEM gatekeeper, 125–126, 131
belief in innate mathematical ability, 128–130
collaborative learning, 133
cultural competence, 131–133
culturally relevant teaching practices, 133–134, 135
community colleges, 134
direct instruction approach, 129
drill and practice pedagogy, 129–130
Eurocentric myth of mathematical knowledge creation, 131
faculty advisor, 132
genetic predisposition, 130, 131–132
high school, 7
math instructor personal history, 123–124, 127–130
online notes, 133
reform, 127, 130–136
rethinking pedagogical approach, 130
seating arrangement, 133
Math illiteracy, 128
Math self-efficacy, 80–81
Mechanical engineer, personal history, 126–127
Mexican American students, STEM persistence, 62
Minorities, *See also* Specific type
associate's degrees, 141–152
data, 145–146

Minorities (*continued*)
  results, 146–150, *147–148*
  study method, 145–146
  community colleges, increased quality and quantity, 104–105
  integrated workforce development systems, underrepresented students, 38, 40, 51–52
  ninth-grade students
    expectancy-value model, 60–72
    STEM persistence, 59–72
  STEM, 156
    acute underrepresentation, 59
    initiatives in community colleges, 144–145
    participation, 40
    trends by discipline, 40
  STEM persistence, 60–72
    domain-specific self-efficacy, 67–68
    framework, 60
    high-ability criteria, 64–65, *65,* 67
    incompatible identities, 63–64
    model, 65, 68
    results, 65–68, *66, 67*
    students' expectations for success, 61–62
    study method, 64–65
    subjective task value, 69–70

N
National Commission on Asian American and Pacific Islander Research in Education, 166
National Science Foundation, 144
  innovation, 43
Ninth-grade students, minorities
  expectancy-value model, 60–72
  STEM persistence, 59–72
Non-cognitive predictors, 76–87
NOVA, 23, 29–30, 31

O
Online notes, math education, 133

P
Pacific Islander students
  articulation agreements, 168
  community colleges
    current initiatives, 163–166
    participation, 159–161
  disaggregated data, 166–167
  ethnic groups, 157–158, 166
  higher education, participation, 158–163
  remediation, 163
  STEM
    current initiatives, 163–166
    participation, 161–163
    student affairs, 167–168
Pedagogy, culturally relevant approaches, 133–136
Pipeline model, STEM, 46
  articulating academic and technical education, 49, 51
  assumption of integrated linearity, 46
  change, 48–49
  conducting sector analysis, 49–50, 51
  integrated career education, 50
  shift from spare parts to system approach, 46–48
  system integration possibilities, 49–50
  toward an integrated system, 48–49
Postsecondary degree, STEM, 23
Postsecondary Education Transcript Study, 8
Practitioner agency, Latinos/as
  actions reflecting agency, 180–183
  assisting Latinos/as to become scientists, 181–182
  barriers to agency, 176–180
  color-blindness, 178–179
  data practices, 177–178
  developing college-knowledge in Latino community, 182
  family-oriented culture, 179–180
  findings, 175–176
  methodology, 174–175, *175, 176*
  one-sided transfer articulation, 177
  program building, 181–183
  self-directed remediation of teaching practices, 182–183
  structural barriers, 177–178
  values and beliefs, 178–180
Propensity score matching
  advantages, 102
  developmental math
    analyses, 113–115, *119–120*
    assessment, 106
    conceptual framework, 110–111
    descriptive results, 113, *119–120*
    environmental factors, 109–110
    independent variables, 110–111
    methodology, 111–115
    placement, 106
    pre-screening, 112–113
    pre-treatment covariates selection, 112–113
    predictors, 116
    study population, 111–112

study purpose, 103
theoretical framework, 109–110
remedial education, 102

R

Race, *See also* Black males; Students of color; Specific type
community colleges
  academic choices, 4–8, *6*
  achievement gap, 97
  adjustment and satisfaction, 94
  analysis, 95–97
  as entry point to postsecondary education, 3
  Beginning Postsecondary Students Longitudinal Study, 4, 8, *9,* 10, *10,* 12, *12*
  bridging research and policy, 13–14
  Education Longitudinal Study, 4, 5, *6*
  family and career responsibilities, 94
  foundational coursework, 11–12, *12*
  four-year STEM majors attending community colleges, 11–12, *12*
  initial choice of STEM major, 8–10, *9*
  intervention, 13
  lack of evidence-based knowledge, 13–14
  learning gains, 96–97, 98
  Postsecondary Education Transcript Study, 8
  previous research results, 93–95
  psychological factors, 91–99
  recent high school graduates, 4–8, *6*
  satisfaction, 96, 97
  STEM transfer, *10,* 10–11
  STEM-related aspirations, 4–8, *6*
  study data source, 95
  toward bachelor's degree, 11–12, *12*
  transfer function, 94
faculty-student interactions, 82–83
interest usage, 84–85
library use, 83–84
role models, 123
STEM, 3–14, 156
  academic choices, 4–8, *6*
  achievement gap, 97
  adjustment and satisfaction, 94
  analysis, 95–97
  Beginning Postsecondary Students Longitudinal Study, 4, 8, *9,* 10, *10,* 12, *12*
  bridging research and policy, 13–14
  Education Longitudinal Study, 4, 5, *6*
  family and career responsibilities, 94

foundational coursework, 11–12, *12*
four-year STEM majors attending community colleges, 11–12, *12*
intervention, 13
lack of evidence-based knowledge, 13–14
learning gains, 96–97, 98
participation trends, 38–42
Postsecondary Education Transcript Study, 8
previous research results, 93–95
psychological factors, 91–99
recent high school graduates, 4–8, *6*
satisfaction, 96, 97
STEM transfer, *10,* 10–11
STEM-related aspirations, 4–8, *6*
study data source, 95
study sample, 95, *96*
toward bachelor's degree, 11–12, *12*
transfer function, 94
underrepresentation, 38
subjective task value, 63
Remedial education, 101–120, *See also* Developmental math
Asian American students, 163
community colleges, 9–10, 102
cost, 101
need for, 101–102
Pacific Islander students, 163
propensity score matching, 102
STEM, 9–10
student outcomes, 107–109
use of, 101–102
Research
  geographical distribution, 44–45
  STEM, promoting, 43–44
Role models
  gender, 123
  race, 123

S

Satisfaction, 96, 97
Scaffolding, STEM, 19
Science, *See also* STEM
  high school, 7
  introductory science courses, 70–71
Science, technology, engineering, and mathematics, *See* STEM
Seating arrangement, math education, 133
Self-efficacy, 61, 79, 85
  defined, 77
Socioeconomic status
  STEM persistence, 65–67
Southeast Asian American students, 163

Index

STEM, 40, 41, *See also* STEM persistence
  alternative analysis to decline, 40–41
  Asian American students
    current initiatives, 163–166
    participation, 161–163
  associate's degrees, 23, 45, 141–152
    data, 145–146
    results, 146–150, *147–148*
    study method, 145–146
  Black males, 193–206
    academic advisors, 201
    academic preparedness, 195–197
    external commitments, 198–200
    faculty support, 198–200
    faculty-student engagement in research opportunities, 200–201
    financial support, 197–198
    future research, 204
    institutional climate, 198–200
    Palmer and Dubord's model of success, *203*
    pedagogical approaches, 198–200
    peer support, 201–204
    policy implications, 204–205
    practice implications, 205–206
    psychological indicators, 198–200
  broadening participation, 44
  certificate, 45
  characterized, 38–39
  community colleges, 3–14, 124–125
    advantages, 22
    change theory, 20–21
    coordination, 28–30
    degree completion, 23
    educators nurturing, 92
    initiatives in, 144–145
    project background, 24
    recent research, 21–33
    state policy impact, 17–33
    state policy implementation, 19–20
    strategic planning, 20–21
    study analysis, 30–33
    study findings, 24–30
    Virginia, 17–33
    Virginia case background, 22–23
  competitiveness, 37
  coordination, 19
  diversity, 3–14
  economic development, 19
  enhancing teaching capacity, 42–43
  ethnicity, 3–14, 156
    academic choices, 4–8, *6*
    achievement gap, 97
    adjustment and satisfaction, 94
    analysis, 95–97
    Beginning Postsecondary Students Longitudinal Study, 4, 8, *9,* 10, *10,* 12, *12*
    bridging research and policy, 13–14
    Education Longitudinal Study, 4, 5, *6*
    family and career responsibilities, 94
    foundational coursework, 11–12, *12*
    four-year STEM majors attending community colleges, 11–12, *12*
    intervention, 13
    lack of evidence-based knowledge, 13–14
    learning gains, 96–97, 98
    participation trends, 38–42
    Postsecondary Education Transcript Study, 8
    previous research results, 93–95
    psychological factors, 91–99
    recent high school graduates, 4–8, *6*
    satisfaction, 96, 97
    STEM transfer, *10,* 10–11
    STEM-related aspirations, 4–8, *6*
    study data source, 95
    study sample, 95, *96*
    toward bachelor's degree, 11–12, *12*
    transfer function, 94
    underrepresentation, 38
  gender
    participation trends, 38–42
    underrepresentation, 38
  geographical distribution, 41
  initiatives for improvement, 42–44
  innovation, promoting, 43–44
  integrated workforce development systems
    need for, 46–50
  underrepresented students, 37–52
  K-12 teaching capacity, 42–43
  Latinos/as, 172–191
    agency study barriers to agency, 176–180
    agency study discussion, 183–189, *184–187*
    agency study findings, 175–176
    agency study methodology, 174–175, *175, 176*
    faculty, administrator or staff agency, 173–191
    out-of-class experiences, 173
    special programs, 172–173
    student agency stance, 173
    what motivates faculty, administrative leaders, and staff, 174

making sense of current initiatives, 44
math as gatekeeper, 125–126, 131
minorities, 156
   acute underrepresentation, 59
   initiatives in community colleges, 144–145
   participation, 40
   trends by discipline, 40
   need for, 59
Pacific Islander students
   current initiatives, 163–166
   participation, 161–163
participation trends, 39–40
pipeline model, 46
   articulating academic and technical education, 49, 51
   assumption of integrated linearity, 46
   change, 48–49
   conducting sector analysis, 49–50, 51
   integrated career education, 50
   shift from spare parts to system approach, 46–48
   system integration possibilities, 49–50
   toward an integrated system, 48–49
postsecondary degree, 23
race, 3–14, 156
   academic choices, 4–8, *6*
   achievement gap, 97
   adjustment and satisfaction, 94
   analysis, 95–97
   Beginning Postsecondary Students Longitudinal Study, 4, 8, *9,* 10, *10,* 12, *12*
   bridging research and policy, 13–14
   Education Longitudinal Study, 4, 5, *6*
   family and career responsibilities, 94
   foundational coursework, 11–12, *12*
   four-year STEM majors attending community colleges, 11–12, *12*
   intervention, 13
   lack of evidence-based knowledge, 13–14
   learning gains, 96–97, 98
   participation trends, 38–42
   Postsecondary Education Transcript Study, 8
   previous research results, 93–95
   psychological factors, 91–99
   recent high school graduates, 4–8, *6*
   satisfaction, 96, 97
   STEM transfer, *10,* 10–11
   STEM-related aspirations, 4–8, *6*
   study data source, 95
   study sample, 95, *96*
   toward bachelor's degree, 11–12, *12*
   transfer function, 94
   underrepresentation, 38
remedial courses, 9–10
research, promoting, 43–44
scaffolding, 19
stereotypes associated, 63
technician preparation, 40, 41
Virginia
   coordination, 28–30
   project background, 24
   study analysis, 30–33
   study findings, 24–30
women
   participation, 40
   trends by discipline, 40
STEM persistence
   Black students, 61–72
      domain-specific self-efficacy, 67–68
      high-ability criteria, 64–65, *65,* 67
      model, 65, 68
      results, 65–68, *66, 67*
      study method, 64–65
      subjective task value, 69–70
   diversity, 59–72
   gender, 65, *66,* 67
      domain-specific self-efficacy, 67–68
      high-ability criteria, 64–65, *65,* 67
      model, 65, 68
      results, 65–68, *66, 67*
      study method, 64–65
      subjective task value, 69–70
   Hispanic students, 61–72
      domain-specific self-efficacy, 67–68
      high-ability criteria, 64–65, *65,* 67
      model, 65, 68
      results, 65–68, *66, 67*
      study method, 64–65
      subjective task value, 69–70
   Mexican American students, 62
   minorities, 60–72
      domain-specific self-efficacy, 67–68
      framework, 60
      high-ability criteria, 64–65, *65,* 67
      incompatible identities, 63–64
      model, 65, 68
      results, 65–68, *66, 67*
      students' expectations for success, 61–62
      study method, 64–65
      subjective task value, 69–70
   research review, 60–61
   socioeconomic status, 65–67
Strategic planning, 20–21

Student affairs
  Asian American students, 167–168
  Pacific Islander students, 167–168
Students of color, community colleges
  action control, 77, 81, 85
  control variables, 81–82
  degree utility, 77, 81, 85
  disparities, 75–76
  English self-efficacy, 81
  faculty-student interactions, 82–83
  field changes to non-STEM majors, 76, *76*
  internet use to access library resources, 84–85
  locus of control, 77, 81, 85
  math self-efficacy, 80–81
  meeting with advisors about academic plans, 83
  non-cognitive predictors, 76–87
  non-cognitive predictors implications, 79, 85–87
  non-cognitive predictors importance, 77–78
  research and practice recommendations, 86–87
  self-efficacy, 77, 85
  STEM, 75–76
  study analytic procedure, 82
  study findings, 82–85
  study methods, 79–80
  study variables, 80–82
  studying at library, 83–84
Subjective task value, 60
  Black students, 62
  culture, 63
  elements, 60
  ethnicity, 63
  expectancy-value theory, 62–63
  intrinsic value component, 62
  race, 63
  utility value component, 62
SySTEMic Solutions, 23, 29–30, 31

T
Teaching capacity
  secondary level, 44, 45

Technician preparation
  community colleges, 45
  high school, 45
  STEM, 40, 41
Technology, *See* STEM
Top Jobs Act
  barriers, 31
  change, 27–28, 31–32, 33
  history, 22–23
  purposes, 17–18
  Virginia, 17–33

U
Underrepresented students, *See* Specific type

V
Virginia
  community colleges
    degree completion, 23
    STEM, 17–33
  STEM
    coordination, 28–30
    project background, 24
    study analysis, 30–33
    study findings, 24–30
  Top Jobs Act, 17–33
Virginia Community College system
  Achieve 2015, 25–27, *26,* 30, 31, 32–33
  context, 24–25
  Dateline 2009, 25
  mission, 24

W
Women
  integrated workforce development systems, underrepresented students, 38, 40, 51–52
  STEM
    participation, 40
    trends by discipline, 40
Workforce development system, three-track system, 47–48
Workforce education
  components, 38
  incoherent system, 38